Hydrogen Production by Water Splitting, Storage and Transportation

This book provides a comprehensive understanding of the process of hydrogen production by water splitting, including materials used, methods and instrumentation. It discusses hydrogen production methods with a focus on water splitting (laboratory/industrial scales) followed by its storage and perspectives. It describes all the methods of hydrogen production, i.e., water electrolysis, steam electrolysis, steam reforming, membrane electrolysis and water splitting. The effects of various radiations (ultraviolet, visible, gamma, X-ray and infrared) on hydrogen production are also included.

Features:

- Presents a complete collection of hydrogen generation and discusses the water splitting process in detail.
- Explores the effects of the radiation of hydrogen generation.
- Discusses hydrogen generation and storage on a large scale.
- Presents a future perspective of hydrogen as fuel.
- Includes future challenges and perspectives to produce hydrogen economically on a large scale.

This book is aimed at graduate students and researchers in materials and chemical engineering, radiation science, physics, chemistry and materials science.

Hydrogen Production by Water Splitting, Storage and Transportation
From Laboratories to Industries

Imran Ali,

Gunel T. Imanova and Al Arsh Basheer

CRC Press

Taylor & Francis Group

Boca Raton London New York

CRC Press is an imprint of the
Taylor & Francis Group, an **informa** business

Designed cover image: Imran Ali

First edition published 2025
by CRC Press
2385 NW Executive Center Drive, Suite 320, Boca Raton FL 33431

and by CRC Press
4 Park Square, Milton Park, Abingdon, Oxon, OX14 4RN

CRC Press is an imprint of Taylor & Francis Group, LLC

ISBN: 9781032458793 (hbk)
ISBN: 9781032558127 (pbk)
ISBN: 9781003432364 (ebk)

DOI: 10.1201/9781003432364

Typeset in Times
by codeMantra

Contents

PART I   Hydrogen production and water splitting

PART II   *Hydrogen storage and transport*

Preface

Discovering sustainable and clean sources is crucial in the ever-changing energy industry. Among the several choices, hydrogen is especially promising as it offers an affordable and eco-friendly replacement for traditional fossil fuels. At the heart of this promise lies the process of water splitting, an incredible chemical event that unlocks hydrogen's potential as a clean energy carrier. This book provides an in-depth look at this fascinating area with ramifications for our energy future as it examines the theories, applications, and possible impacts of water splitting on hydrogen production. Readers are guided through the intricate dynamics and transformative potential of this process in each chapter, which covers everything from fundamental chemistry to cutting-edge technologies. We commence our investigation by delving into the fundamental concepts of water splitting, clarifying the molecular nuances that control its practicability and effectiveness. Next, we explore the wide range of catalytic strategies used for this reaction, ranging from conventional electrolysis to recently developed photovoltaic and photo-electrochemical approaches. This book seeks to offer a thorough understanding of hydrogen production by water splitting via the lenses of research, innovation, and practical application. We sincerely hope that this investigation will stimulate scholars, decision-makers, and business executives to quicken the shift to a more sustainable and cleaner energy future. We sincerely believe that this book will serve as a valuable resource for future research, scientists, and business professionals involved in the generation of green energy.

Acknowledgments

It was indeed a difficult task for us to complete this book but the extreme help and co-operation of our research scholars and other laboratory staff made this job easy. Finally, Taylor and Francis is also acknowledged for giving an opportunity to write this book.

Imran Ali

Gunel Imanova

Al Arsh Basheer

Author Biographies

Imran Ali

Imran Ali is currently an academician and researcher at the Department of Chemistry, Jamia Millia Islamia (Central University), Jamia Nagar, New Delhi – 110025, India. Prof. Imran Ali, Ph.D., FRSC, C Chem, London (UK), Highly Cited Researcher, Clarivate, USA, and first rank in India and tenth rank in the world in analytical chemistry, as per the Stanford University, USA (Global list of top 2% scientists), is a world-recognized academician and researcher. He completed his Ph.D. at the Indian Institute of Technology Roorkee, Roorkee, India. He is known globally due to his great contribution to multidisciplinary research areas dedicated to water treatment, hydrogen production by water splitting, separation science, and the development of nano-anticancer drugs. He has published more than 550 papers in reputed journals. He has also written six books published by reputed publishers. His total citation is 42,500 with an h-index of 111 and i10-index of 364. He is a member of various scientific societies globally. He has been co-chair of a conference on the application of graphene, chaired many conference sessions, and delivered several keynote lectures. He is a widely traveled person enjoying various visiting professor/consultant positions in many universities around the world. He has research collaborations in more than ten countries all over the world.

Gunel Imanova

Gunel T. Imanova is currently working as a leading researcher at the Institute of Radiation Problems, Ministry of Science and Education Republic of Azerbaijan. She is a scientific researcher and lecturer at Khazar University (Department of Physics and Electronics). She is a scientific researcher at Western Caspian University (Azerbaijan) and Azerbaijan State University of Economics (UNEC Research Center for Sustainable development and Creen economy named after Nizami Ganjavi). She graduated from Baku State University with a red diploma (distinction), a bachelor's degree and a master's degree. In 2018, she was awarded a Ph.D. degree in physics. In 2022, she was awarded with the title of associate professor. Due to her scientific productive activities in 2018 and 2022, she was awarded diplomas by the Institute of Radiation Problems, Ministry of Science and Education Republic of Azerbaijan, Council of Young Scientists and Specialists, Trade Union Organization, and Women's Council. She has participated in many international conferences in 2020–2023 and was awarded more than 100 participant and speaker certificates. She is continuing her scientific research on the topic of Doctor of Science at the institute where she is currently working. She is a member of the editorial board of 16 foreign publishing houses (in one of these memberships, she is the editor-in-chief of a journal published in Azerbaijan (Caspian Journal of Energy)). She is the author of more than 100 scientific works. She has published three books i.e. one monograph in Azerbaijan and two books in Moldova. Her total citation is 800 with an h-index of 17 and an i10-index of 35. Presently, she is working in the area of hydrogen production by water splitting using gamma radiation and a variety of catalysts. She has research collaborations in more than ten countries all over the world.

Al Arsh Basheer

Al Arsh Basheer is a young and dynamic researcher. He obtained his bachelor's degree in mechanical and aerospace engineering (Double major) from the University at Buffalo, The State University of New York, USA. He also obtained his master's degree in robotics from the University of California, Davis, USA. He has published 16 articles in peer-reviewed journals with 3100 total citations and an h-index of 12. The importance of his work is reflected in the fact that some of his papers are cited more than 500 times within five years' time. He is on the editorial board of some journals. He is also a member of some societies like the Royal Society of Chemistry, the American Institute of Aeronautics and Astronautics, and the American Society of Mechanical Engineers at the University of Buffalo. He is continuously doing research in his leisure on the topics of hydrogen generation, environmental science, and chirality.

Part I

Hydrogen production and water splitting

1 Introduction

1.1 GENERAL INTRODUCTION

Energy plays a crucial role in the world's economy, environment, and geopolitics. The development and expansion of the economy mostly depend on energy. The energy is required for industries, transport, homes, businesses, and various other activities. The productivity in the economy depends on having reliable, reasonably priced energy. The world uses a variety of energy sources, such as nuclear energy; renewable energy sources including solar, wind, hydro, and geothermal energy; and fossil fuels like coal, oil, and natural gas. The mix's composition can have a big impact on environmental sustainability and global energy security. The amount of energy consumed varies throughout various nations and locations. Developed countries tend to consume more energy per person on average, whereas emerging economies are using more energy as they expand. A global challenge is striking a balance between sustainability and energy access. The production and consumption of energy around the world have a big impact on the environment. The emissions of greenhouse gases are a result of the use of fossil fuels. In order to combat climate change, switching to cleaner, renewable energy sources is essential. Energy resource control and access have always been central to geopolitical disputes. Diversifying energy sources and developing relationships with energy-rich nations are common ways that nations strive for energy security [1,2].

The coordination and facilitation of international collaboration on energy-related issues are the tasks of a number of organizations, including the International Energy Agency and the International Renewable Energy Agency. A steady and dependable energy source is a challenge for all nations. This entails securing vital energy infrastructure, minimizing reliance on a single energy provider, and maintaining a variety of energy sources. One of the main focuses of the Sustainable Development Goals of the United Nations is sustainable energy. It is connected to objectives including eradicating poverty, protecting the environment, and providing access to essential services. A number of global issues, such as security, economic growth, and climate change, are closely related to energy. Energy is intricately linked to various global challenges, including climate change, economic development, and security. The world should address the increasing energy needs and environmental concerns, i.e., how we create, use, and think about energy. This will have a significant impact on the planet's development in the future. Around the world, a large number of people, particularly in developing nations, lack access to cheap and dependable electricity. One of the main objectives of global development is to increase access to clean and sustainable energy sources [3]. The enhancement of energy efficiency is a top priority worldwide. It is possible to meet energy demand while minimizing environmental effects by reducing energy waste through the use of more efficient technology and practices. The global push for an energy transition involves shifting from fossil fuels

DOI: 10.1201/ 9781003432364-2

to cleaner energy sources. This transition is driven by concerns about climate change and air pollution, as well as technological advancements and market forces [4].

1.2 NEED FOR GREEN ENERGY FOR CLIMATE ISSUES

In order to solve climate challenges, green or sustainable energy sources are essential. Here are a few main reasons why using green energy is crucial to reducing climate change and the problems it causes. When fossil fuels like coal, oil, and natural gas are used to produce energy, they release greenhouse gases like carbon dioxide (CO_2) and methane (CH_4) into the atmosphere. The main causes of climate change and global warming are these emissions. Green energy sources help to lower the global carbon footprint since they provide electricity with little to no direct greenhouse gas emissions, such as solar, wind, hydro, and geothermal power. The rising sea levels, more frequent and intense heatwaves, storms, and ecological disturbances are just a few of the issues linked to climate change brought on by the buildup of greenhouse gases in the atmosphere. Making the switch to renewable energy sources is essential to reducing global warming and averting its worst effects. Compared to conventional fossil fuel-based systems, many green energy methods are intrinsically more energy-efficient. For instance, light-emitting diodes are more energy-efficient than incandescent bulbs, and electric vehicles are more energy-efficient than cars with internal combustion engines. The emissions are decreased, and the overall energy demand is decreased with increased energy efficiency [5–7].

Green energy sources are frequently linked to more environmentally friendly performances. They promote water conservation, prudent land use, and less pollution of the air and water. This can lead to a healthier ecosystem and environment. Green energy sources are frequently domestically available, reducing a nation's dependence on energy imports. This can improve energy security by spreading energy sources and reducing exposure to international struggles over energy resources. The development and deployment of green energy technologies arouse innovation and generate economic chances. This can lead to the development of clean energy trades and the make of green jobs. International agreements, such as the Paris Agreement, set goals for dropping greenhouse gas emissions. The transition to green energy is vital for meeting these assurances and evading disastrous climate change [8,9].

The burning of fossil fuels not only contributes to climate alteration but also results in air contamination, which can have serious health effects. Shifting to green energy can increase air quality and public health by reducing harmful emissions like particulate matter and sulfur dioxide. Distributed renewable energy systems, such as rooftop solar panels, can allow individuals and communities to create their clean energy and decrease their dependence on central power grids. Climate alteration poses a noteworthy danger to biodiversity, as shifts in temperature and weather styles can disturb ecosystems. Green energy plays a role in protecting biodiversity and preserving ecosystems by reducing greenhouse gas emissions [10,11]. Briefly, the transition to sustainable or green energy sources is dynamic for fighting climate change, dropping greenhouse gas emissions, and addressing a wide range of environmental and social challenges. It represents a critical step toward a more sustainable and climate-resilient future.

TABLE 1.1

Different hydrogen colors, methods of preparation, price, and CO$_2$ emission [12]

H$_2$ color	Production methods	CO$_2$ emissions	Price
Blue	Natural gas steam reforming, with carbon capture and storage	9–12 kg CO$_2$/kg H$_2$	Moderate
Brown	Coal gasification or other feedstock	19–25 kg CO$_2$/kg H$_2$	Moderate
Gray	Natural gas steam reforming, without carbon capture and storage	16–18 kg CO$_2$/kg H$_2$	Low
Black	Gasification of coal with hydrogen separation and release of other gases	24–28 kg CO$_2$/kg H$_2$	High
Green	Water splitting using renewable energy sources	Zero CO$_2$ emissions	High

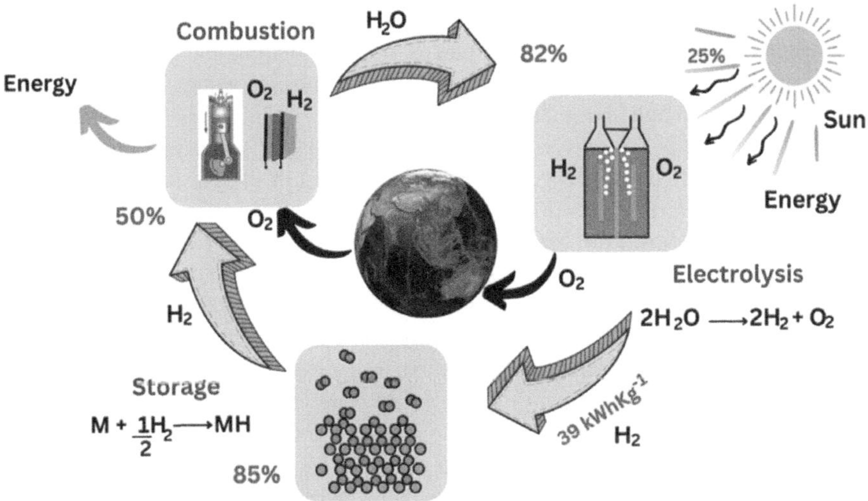

FIGURE 1.1 Representation of hydrogen cycle.

In view of these facts, scientists have started to explore the green and alternative sources of energy. The various forms of alternative sources of energy are solar, wind, hydropower, geothermal energy, biomass energy, tidal wave energy, nuclear energy, ocean thermal energy, waste-to-energy, and hydrogen. Among them hydrogen is considered the best one due to its ease of preparation, high energy value (120–142 MJ/kg heat value), and no toxicity. Hydrogen occurs in blue, gray, brown, black, and green forms depending on the source of production. Blue hydrogen is created by steam reforming of natural gas, with carbon capture and storage. Gray hydrogen is created from steam reforming of natural gas, without carbon capture and storage. Brown hydrogen is created from the gasification of coal or other hydrocarbon-rich feedstock. Black hydrogen is created from coal gasification, with hydrogen separation and other gases released into the atmosphere. Green hydrogen is produced using renewable energy sources, such as wind and solar power by water splitting. The various forms of hydrogen along with cost and CO_2 emission are shown in Table 1.1 [12]. Among these categories, green hydrogen is assumed as the cleanest and most environmentally friendly form of hydrogen, as it creates no direct greenhouse gas emissions during production. The use of hydrogen for energy needs contains three major steps: (i) production, (ii) storage, and (iii) application through combustion. These three major steps constitute a closed hydrogen cycle, as shown in Figure 1.1 [13]. The important methods include thermochemical, steam reforming, electrolysis, photo-electrochemical (PEC), and photocatalytic (PC). This book is dedicated to hydrogen production by water splitting through PEC and PC methods.

1.3 IMPORTANCE OF GREEN HYDROGEN ENERGY FUEL

The chemical element hydrogen has the atomic number one and the symbol H. This is the lightest element in the periodic table. Hydrogen is a gas composed of diatomic molecules (H_2) under normal circumstances. It is an inflammable gas but not a supporter of combustion. It is lighter than air and insoluble in water. It has an atomic mass of 1.008 amu and an ionization enthalpy of 1,312 kJ/mol. It is the most copious chemical material in the universe, making roughly 75% of all normal matter. Many stars including Sun are chiefly composed of hydrogen in the plasma state. Most of the hydrogen on Earth exists in molecular forms like water and organic substances. Hydrogen is being used for various applications with a major role as green energy [14]. Green energy hydrogen fuel, often called green hydrogen, has notable importance for many key reasons. It has a high heat value of 120–142 MJ/kg compared to 50–55, 29.0, and 22.7 MJ/kg for methane, dimethyl ether, and methanol, respectively. The other properties of hydrogen with comparison are summarized in Table 1.2 [12].

Green hydrogen is created by electrolyzing water to separate its hydrogen and oxygen ions using renewable energy sources like solar and wind power. There are no direct greenhouse gas emissions from this technique. Green hydrogen is, therefore, a clean, low-carbon energy source that helps in the fight against climate change by lowering carbon emissions. Green hydrogen has the potential to be an energy storage technology. When energy demand is high or renewable energy generation is low, it enables excess renewable energy produced during peak production periods to be stored as hydrogen and subsequently consumed. This offers a way to balance the intermittent nature of renewable energy sources and aids in system stabilization. Green hydrogen is a flexible energy source with a broad range of uses. It can be applied to a number of industries, including industrial operations, stationary power generation, and transportation. For example, it can power hydrogen-fueled cars, supply heat and energy to stationary fuel cells, and take the place of fossil fuels in industrial processes like the manufacturing of ammonia. Local production of green hydrogen from renewable energy sources can lessen reliance on imports of fossil fuels and improve energy security. For areas that are dependent on energy imports but have abundant renewable resources, this can be especially advantageous. Air quality can be enhanced by substituting renewable hydrogen for fossil fuels in a variety of

TABLE 1.2

Physical properties of the hydrogen molecule with comparison of other gases

Properties	Hydrogen	Other gases
Density (gas)	0.083 kg/m^3 (0°C, 1 bar)	~1/10 of natural gas
Density (liq.)	70.85 kg/m^3 (−253°C, 1 bar)	~1/6 of natural gas
Boiling point	−252.76°C (1 bar)	~90°C below LNG (liquid natural gas)
Energy density (ambient cond., lower heating value)	1.10 × 10^4 J/L	~1/3 of natural gas
Energy per unit of mass (lower heating value)	1.18 × 10^8 J/kg	~3× that of gasoline
Velocity of flame	1.7 m/s	~8× methane
Specific energy (liquefied, lower heating value)	8.4 × 10^6 J/L	~1/3 of LNG
Range of ignition	4–75% in air by volume	~6× wider than methane
Energy of ignition	2.1 × 10^4 J	~1/10 of methane
Temperature of auto-ignition	535°C	280°C for gasoline

applications. Green hydrogen does not release any toxic emissions that could affect the environment or human health, such as sulfur dioxide, nitrogen oxides, or particulate matter. The preparation of green hydrogen technologies and industries can rouse economic growth and generate jobs in hydrogen production, supply chains, and application industries. This can aid in diversifying economies and support a change-over to a sustainable energy future [15].

It can also be transported as safe energy when converted into a carrier such as ammonia, a zero-carbon fuel for shipping. The search for green hydrogen has driven technological signs of progress in electrolysis, renewable energy integration, and hydrogen storage and transportation. These innovations have broader applications beyond green hydrogen production. Green hydrogen can power hydrogen fuel cell vehicles (FCVs), which can contribute to transportation that is carbon neutral. These vehicles release only water vapor, making them a clean and sustainable alternative to internal combustion engine vehicles. The green hydrogen is measured as a long-term and sustainable energy solution. The green hydrogen is flexible and can be transformed into electricity or synthetic gas and utilized for industrial, commercial, or mobility purposes. As renewable energy capability carries on to grow and the costs of green hydrogen production reduce, it has the possibility to become a reliable and sustainable source of energy. The production and distribution of green hydrogen can nurture international cooperation and businesses. The regions with abundant renewable energy resources can export green hydrogen to areas with high energy needs but restricted renewable capability [16–19]. Briefly, the green energy hydrogen fuel is important for addressing climate change, improving air quality, reducing carbon emissions, and fostering sustainable energy systems. Its adaptability, prospective for energy storage, and role in multiple sectors make it a valued component of the transition to a cleaner and more sustainable energy future.

1.4 CURRENT STATUS OF HYDROGEN AS A GREEN FUEL

The current position of hydrogen as a green fuel is still inadequate, and most hydrogen produced globally is gray. The green hydrogen is still in its initial stages of development but it is a critical solution to decarburization and reducing our dependence on fossil fuels. Also, hydrogen as a green fuel is attaining noteworthy attention and impetus. There may have been further developments and variations in the position of hydrogen as a green fuel. Presently, little hydrogen is green because the procedure involved in making green hydrogen is enormously energy-intensive. These methods are not low cost and need a high amount of energy. The cost of green hydrogen varies from 3 to 8 euros per kg in different areas of the world. The price of green hydrogen is higher than gray hydrogen. Nevertheless, the price of green hydrogen is predicted to reduce as technology advances and renewable energy becomes more widely accessible [20,21].

The status of hydrogen is observed, and it is seen that electrolysis, a process that utilizes electricity from renewable sources to split water into hydrogen and oxygen, is the main method for producing green hydrogen. The electrolyzer producers are scaling up the manufacturing, and costs are reducing, making green hydrogen more inexpensive. The electrolysis units are being arranged near renewable energy to capture excess energy and convert it into hydrogen for storage and utilization. Hydrogen FCVs exist in some markets, and several automakers are working to expand their FCV contributions. Hydrogen-powered buses and trucks are being organized for public transportation and commercial uses. The fuel cells utilizing green hydrogen are being applied for stationary power generation in some areas. These applications included distributed energy generation, backup power systems, and powering critical infrastructure. The industries such as steel, chemical manufacturing, and ammonia production are exploring the use of green hydrogen to decarbonize their processes. Some projects intended to replace fossil fuels with green hydrogen in high-temperature industrial applications are in progress [22,23].

The efforts are ongoing to develop hydrogen setup, including storage facilities, refueling stations, and transportation networks. The investment in infrastructure is important for allowing the widespread acceptance of hydrogen as a fuel. The environmental and energy policies of many nations are progressively focusing on dipping carbon emissions and encouraging green hydrogen as a means to achieve these objectives. Various countries and regions are working on joint projects and enterprises to encourage the use of green hydrogen. International organizations and agreements are also supporting hydrogen as a clean energy carrier. Governments, energy companies, and research institutions are investing seriously in hydrogen-related research and projects. Many countries have declared hydrogen strategies and funding initiatives to support the development of green hydrogen. The most important countries are Austria, Australia, Belgium, Brazil, China, European Union, France, Germany, Italy, India, Japan, New Zealand, Norway, South Korea, Saudi Arabia, South Africa, the Netherlands, United Kingdom, and the United States of America [24]. The integration of renewable energy sources, such as wind and solar power, into hydrogen production is a key attention [25–28].

At the end of 2023, many hydrogen refueling stations were working globally. During the write-up of this book, it was realized that hydrogen can be filled up in

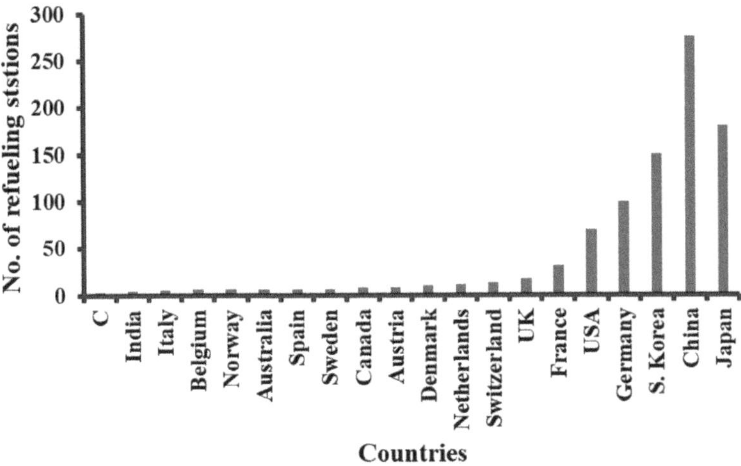

FIGURE 1.2 Hydrogen refueling stations at the end of 2023.

C: Hungary, Iceland, Croatia, Slovenia, Costa Rica, Czech Republic, Latvia, Malaysia, Saudi Arabia, Taiwan, and the UAE.

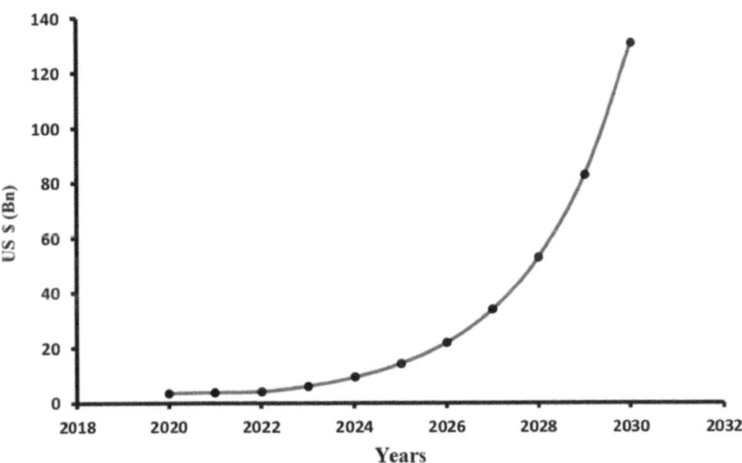

FIGURE 1.3 Status of green hydrogen market with future prediction.

about 30 countries. The total number of hydrogen refueling stations in various countries in 2023 years is counted and given in Figure 1.2.

The green hydrogen is expected to increase continuously. The current market status in 2023 of green hydrogen is 6.10 billion US$. This market will increase in the future and the expected rise may go up to 13.1 billion US$ in 20230 as shown in Figure 1.3.

Hydrogen production may go up to 225 metric tons in 2050. This assumption is based on our past experience, past green hydrogen production, and the reading of

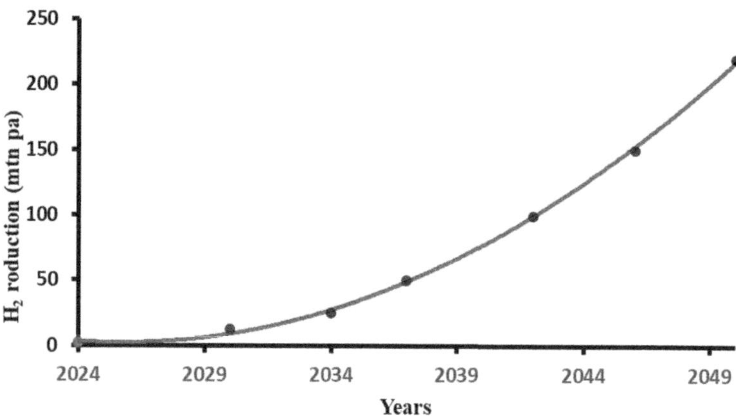

FIGURE 1.4 Predicted global demand for green hydrogen.

various papers and books during the write-up of this book. This is due to the fact that the global demand for hydrogen is increasing continuously. The data were collected and a graph as shown in Figure 1.4 was prepared. This figure clearly indicates a growing demand for green hydrogen.

REFERENCES

1. K. Mehta, M. Ehrenwirth, W. Zörner, R. Greenoug, Need of energy transition at roof of the world: Correlative approach to interpret energy identity of high-altitude Central Asian communities, *Energy Sustain. Dev.*, 76, 101271 (2023).
2. O.O. Yolcan, World energy outlook and state of renewable energy: 10-Year evaluation, *Innovation Green Dev.*, 2, 100070 (2023).
3. H.H. Pourasl, R.V. Barenji, V.M. Khojastehnezhad, Solar energy status in the world: A comprehensive review, *Energy Rep.*, 10, 3474–3493 (2023).
4. C. Mertzanis, Energy policy diversity and green bond issuance around the world, *Energy Econ.*, 128, 107116 (2023).
5. A.G. Olabi, M.A. Abdelkareem, Renewable energy and climate change, *Renew. Sustain. Energy Rev.*, 158, 112111 (2022).
6. A. Pani, S.S. Shirkole, A.S. Mujumdar, Importance of renewable energy in the fight against global climate change, *Dying Technol.*, 40, 2581–2582 (2022).
7. E. Jamei, H.W. Chau, M. Seyedmahmoudian, S. Mekhilef, F.S. Hafez, Green roof and energy – role of climate and design elements in hot and temperate climates, *Heliyon*, 9, e15917 (2023).
8. P.A. Owusu1, S. Asumadu-Sarkodie, A review of renewable energy sources, sustainability issues and climate change mitigation, *Cogent Eng.*, 3, 1167990 (2016).
9. A.I. Osman, L. Chen, M. Yang, G. Msigwa, M. Farghali, S. Fawzy, D.W. Rooney, P.-S. Yap, Cost, environmental impact, and resilience of renewable energy under a changing climate: A review, *Environ. Chem. Lett.*, 21, 741–764 (2023).
10. V.L. Kalyani, M.K. Dudy, S. Pareek, Green energy: The Need of the world, *J. Manag. Eng. Inf. Technol.*, 2, 18–26 (2015).
11. P.F. Borowski, Mitigating climate change and the development of green energy versus a return to fossil fuels due to the energy crisis in 2022, *Energies*, 15, 928 (2022).

12. I. Marouani, T. Guesmi, B.M. Alshammari, K. Alqunun, A. Alzamil, M. Alturki, H.H. Abdallah, Integration of renewable-energy-based green hydrogen into the energy future, *Processes*, 11, 2685 (2023).

13. A. Zuttel, A. Remhof, A. Borgschulte, Hydrogen: The future energy carrier, *Philos. Trans. R. Soc. A*, 368, 3329–3342 (2010).

14. P. Hota, A. Das, D.K. Maiti, A short review on generation of green fuel hydrogen through water splitting, *Int. J. Hydrogen Energy*, 48, 523–541 (2023).

15. U.S. Meda, Y.V. Rajyaguru, A. Pandey, Generation of green hydrogen using self-sustained regenerative fuel cells: Opportunities and challenges, *Int. J. Hydrogen Energy*, 48, 28289–28314 (2023).

16. S.M.M. Ehteshami, S.H. Chan, The role of hydrogen and fuel cells to store renewable energy in the future energy network – potentials and challenges, *Energy Policy*, 73, 103–109 (2014).

17. U. Sontakke, S. Jaju, Green hydrogen economy and opportunities for India, *IOP Conf. Ser.: Mater. Sci. Eng.*, 1206, 012005 (2021).

18. M. Pérez-Vigueras, R. Sotelo-Boyás, R.D.G. González-Huerta, F. Bañuelos-Ruedas, Feasibility analysis of green hydrogen production from oceanic energy, *Heylion*, 9, e20046 (2023).

19. A.G. Olabi, M.A. Abdelkareem, M.S. Mahmoud, K. Elsaid, K. Obaideen, H. Rezk, T. Wilberforce, T. Eisa, K.-J. Chae, E.T. Sayed, Green hydrogen: Pathways, roadmap, and role in achieving sustainable development goals, *Process Saf. Environ. Prot.*, 177, 664–687 (2023).

20. Y. Manoharan, S.E. Hosseini, B. Butler, H. Alzhahrani, B.T.F. Senior, T. Ashuri, J. Krohn, Hydrogen fuel cell vehicles; Current status and future prospect, *Appl. Sci.*, 9, 2296 (2019).

21. P.-A. Le, V.D. Trung, P.L. Nguyen, T.V.B. Phung, J. Natsukicd, T. Natsuki, The current status of hydrogen energy: An overview, *RSC Adv.*, 13, 28262 (2023).

22. E.B. Agyekum, C. Nutakor, A.M. Agwa, S. Kamel, A critical review of renewable hydrogen production methods: Factors affecting their scale-up and its role in future energy generation, *Membranes*, 12, 173 (2022).

23. T. Jamal, G.M. Shafiullah, F. Dawood, A. Kaur, M.T. Arif, R. Pugazhendhi, R.M. Elavarasan, S.F. Ahmed, Fuelling the future: An in-depth review of recent trends, challenges and opportunities of hydrogen fuel cell for a sustainable hydrogen economy, *Energy Rep.*, 10, 2103–2127 (2023).

24. The Future of Hydrogen: Seizing today's opportunities, Report prepared by the IEA for the G20, Japan (2019).

25. M. Younas, S. Shafique, A. Hafeez, F. Javed, F. Rehman, An overview of hydrogen production: Current status, potential, and challenges, *Fuel*, 316, 123317 (2022).

26. J. Ni, Y. Wen, D. Pan, J. Bai, B. Zhou, S. Zhao, Z. Wang, Y. Liu, Q. Zeng, Light-driven simultaneous water purification and green energy production by photocatalytic fuel cell: A comprehensive review on current status, challenges, and perspectives, *Chem. Eng. J.*, 473, 145162 (2023).

27. C. Kim, S.H. Cho, S.M. Cho, Y. Na, S. Kim, D.K. Kim, Review of hydrogen infrastructure: The current status and roll-out strategy, *Int. J. Hydrogen Energy*, 48, 1701–1716 (2023).

28. S. Harichandan, S.K. Kar, P.K. Rai, A systematic and critical review of green hydrogen economy in India, *Int. J. Hydrogen Energy*, 48, 31425–31442 (2023).

2 Methods of hydrogen production

2.1 INTRODUCTION

It is important to keep in mind that different hydrogen production techniques have different environmental effects and energy efficiency levels. This depends on the specific technology used and the types of energy used, i.e., fossil fuels or renewable sources. One of the main goals for developing the hydrogen economy is to switch to low-carbon and sustainable green energy techniques. The best method of hydrogen production involves low-cost paraphernalia and raw materials. Also, the method steps should be as minimal as possible. The methods should be human and environmentally friendly. Hydrogen is produced from water by water splitting using different techniques. The important methods include thermochemical, steam reforming, electrolysis, photo-electrochemical, and photocatalytic (PC). A brief comparison of these methods for hydrogen production is given in Table 2.1. However, this book is dedicated to hydrogen production by water splitting through photo-electrochemical (PEC), and PC methods.

TABLE 2.1
A comparison of various methods for hydrogen production using water as feed

Methods	Sources of energy	Advantages	Challenges
Thermochemical	Thermal	Large-scale production, utilization of heat from waste materials	Requires heat-resistant materials, slow heating-cooling cycles
Steam reforming	Thermal	Inexpensive, good production, commercialized	Green gas emissions and fissile fuels are deployed
Electrolysis	Electricity	No green gas emissions, integration with renewable energy	Low efficiency, difficulty in scaling up
Photo-electrochemical	Electricity and radiation	No green gas emissions, integration with renewable energy	Low efficiency, difficulty in scaling up
Photocatalytic	Radiation	No green gas emissions, integration with renewable energy	Low efficiency, difficulty in scaling up, under-developing stage

DOI: 10.1201/9781003432364-3

2.2 STEAM REFORMING

Steam reforming for water splitting is a conventional method for hydrogen generation by using a mixture of two different procedures, i.e., steam methane reforming (SMR) and water splitting. In this process, methane (from natural gas) is heated with steam in the presence of a catalyst at 3–25 bar pressure. It produces a mixture of carbon monoxide and hydrogen with some amount of carbon dioxide—a greenhouse gas [1–6]. Both of these methods can produce hydrogen but have different purposes and uses [7]. The SMR is an extensively used industrial process for hydrogen manufacture. It comprises the reaction of methane (CH_4), naturally attained from natural gas, with steam (H_2O) at high temperatures (usually 700°C–1,000°C) using a catalyst [8].

The chief chemical reaction is:

$$CH_4 + H_2O \rightarrow CO + 3H_2$$

This reaction yields carbon monoxide (CO) and hydrogen gas (H_2). The manufactured hydrogen can be utilized for numerous applications, including fuel cells and industrial procedures.

Water splitting is a procedure that utilizes electrical energy to break water (H_2O) into hydrogen (H_2) and oxygen (O_2) by passing an electric current in an electrolyte. There two are half-reactions involved as shown below.

$$\text{Anode (oxidation): } 2H_2O \rightarrow O_2 + 4H^+ + 4e^- \tag{2.1}$$

$$\text{Cathode (reduction): } 4H^+ + 4e^- \rightarrow 2H_2 \tag{2.2}$$

Briefly, steam reforming and water splitting imply a mixture of SMR and water splitting via electrolysis. Each of these processes is normally carried out separately for their intended purposes [9]. The significant endothermic reaction and constant heat input to the process lead to an increasing heat load, which is the drawback. The schematic representation of steam reforming is shown in Figure 2.1.

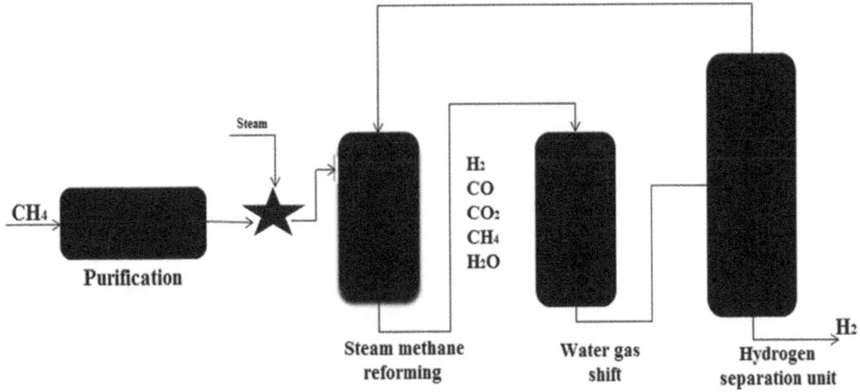

FIGURE 2.1 A schematic representation of steam reforming.

2.3 THERMOCHEMICAL

Thermochemical water splitting is a procedure that uses high-temperature chemical reactions to break water (H_2O) into its elements, i.e., hydrogen (H_2) and oxygen (O_2). This is a substitute method for hydrogen production. This has the prospective to be more effective and ecologically friendly compared to outdated methods like SMR. Thermochemical water splitting usually includes a sequence of chemical reactions that happen in a cycle, and the cycle can be recurrent to endlessly produce hydrogen. There are numerous thermochemical water-splitting cycles, and two of the most well-known ones are the sulfur-iodine cycle (S-I cycle) and the cerium-based cycle [10–12].

S-I Cycle: This cycle involves sulfur dioxide (SO_2) reaction with water to get sulfuric acid (H_2SO_4). Then, the sulfuric acid is disintegrated into sulfur trioxide (SO_3) and water. The sulfur trioxide is reacted with hydrogen to form sulfur dioxide and water, finishing the loop. This cycle can be utilized to make hydrogen, but it needs high temperatures and is still in the research and development stage.

Cerium-Based Cycle: This cycle includes the use of cerium oxide (CeO_2) as a redox material. It functions at very high temperatures. Cerium oxide is reduced by high-temperature heat to yield CeO_{2-x} (where x is less than 2) and oxygen. The reduced cerium oxide can then react with water vapors to yield hydrogen and regenerate cerium oxide. The cycle is recurring by providing heat to the cerium oxide. A schematic representation of hydrogen production by cerium-based cycle is shown in Figure 2.2.

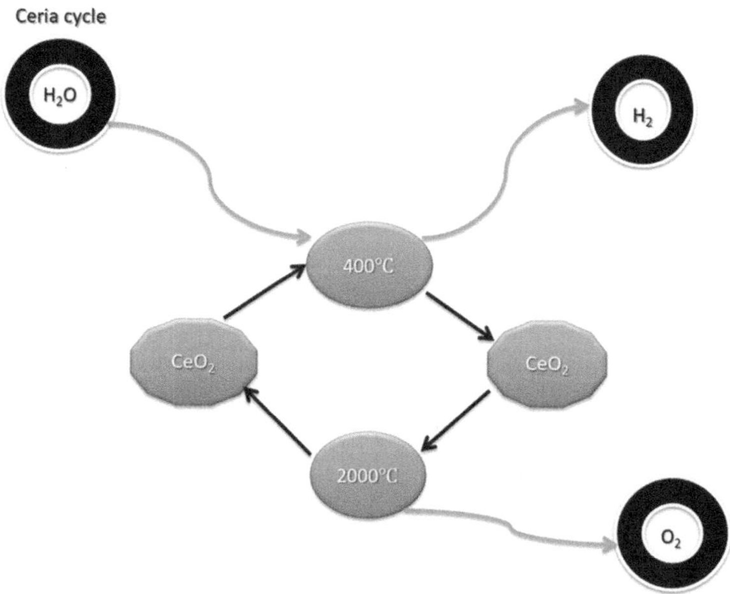

FIGURE 2.2 A schematic representation of hydrogen production by cerium-based cycle.

The advantages of thermochemical water splitting for hydrogen generation are given below:

- High temperature is obtained by many sources like solar, nuclear, etc. sources.
- Ability to achieve higher thermodynamic efficiencies compared to low-temperature.
- The potential for on-demand hydrogen production.

The challenges and limitations of this technology are given below:

- High temperatures can lead to material degradation and heat management challenges.
- Complex reaction cycles and the need for efficient catalysts.
- High capital and operating costs owing to the extreme conditions required.
- Limited industrialization and scalability compared to other hydrogen production methods.

The research in thermochemical water splitting is continuing, and progressions are being made to increase the feasibility and efficiency of this hydrogen production method. It has the prospect to play a role in a future hydrogen economy, particularly in applications where high-temperature heat sources are freely available and emissions need to be minimalized.

2.4 ELECTROLYSIS

Electrolysis is the phenomenon of water splitting into hydrogen and oxygen using electricity. This reaction takes place in a special vessel called an electrolyzer. In this phenomenon, electrical energy is converted into chemical energy. This technique is being used to split water for hydrogen and oxygen. The electrolysis is of different types and some of the modes of electrolysis for water splitting are discussed below.

2.4.1 NORMAL WATER ELECTROLYSIS

The primary goals of water electrolysis are the production of clean oxygen and hydrogen gasses. The water is subjected to an electric current, which causes the water to break down into hydrogen and oxygen. One of the most effective ways to manufacture hydrogen is by electrolysis of water, which uses cheap water and yields only pure oxygen as a byproduct. A basic setup for demonstrating water electrolysis at home is shown in Figure 2.3.

DC power from renewable energy sources, such as biomass, solar, and wind, is used in the electrolysis process. One of its benefits is that it can create hydrogen as renewable energy. In order to boost the usage of water electrolysis, existing electrolyzers must be made more efficient, durable, and safe while also using less energy, money, and maintenance. The addition of salts can speed up any electrolysis of water experiment because water itself does not conduct electricity. In addition, the energy

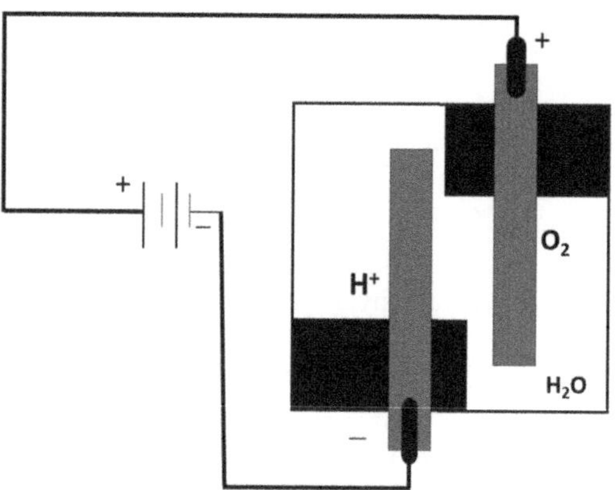

FIGURE 2.3 A simple set-up for the demonstration of electrolysis of water at home.

required to create oxygen and hydrogen is significant. However, burning fossil fuels to generate the energy needed for electrolysis water results in significant CO_2 emissions from the process [13–17].

2.4.2 Steam Electrolysis

It is the method of water electrolysis at high temperatures. High-temperature electrolysis (HTE) is a technology for producing hydrogen from water at high temperatures. HTE, also known as steam electrolysis, is the process of splitting water to produce oxygen (O_2) and hydrogen (H_2) at temperatures between 700°C and 1,000°C. Electrical energy is the driving force behind this process. It has been proven that electrolysis of water vapor in solid-oxide electrolyte cells is a very effective way to produce hydrogen from water. HTE is shown in Figure 2.4.

In HTE, water is first changed to steam by means of nuclear thermal energy rather than electricity, and then separated at the cathode to create hydrogen and oxygen ions. The oxygen ions afterward travel through the solid oxide electrolyte material, and then create oxygen molecules at the anode surface [18–21]. The major drawback of this method is the improved heat load resulting from the big endothermic reaction and the constant heat supply to the reaction. Also, some green gases like carbon dioxide and carbon monoxide are produced—a concern of global warming and climate change.

2.4.3 Membrane Electrolysis

The membrane electrolysis is carried out by proton exchange membranes (PEMs). Proton exchange membranes, also known as polymer electrolyte membranes or PEMs, are semipermeable membranes that are often composed of monomers and are intended to transmit protons while serving as reactant barriers, such as those to oxygen and hydrogen gas, and electrical insulators. The generated protons' conduction from the anode to the cathode is the function of the membrane that sits between electrodes. Through the electrolyte membrane, the existing ions travel toward the

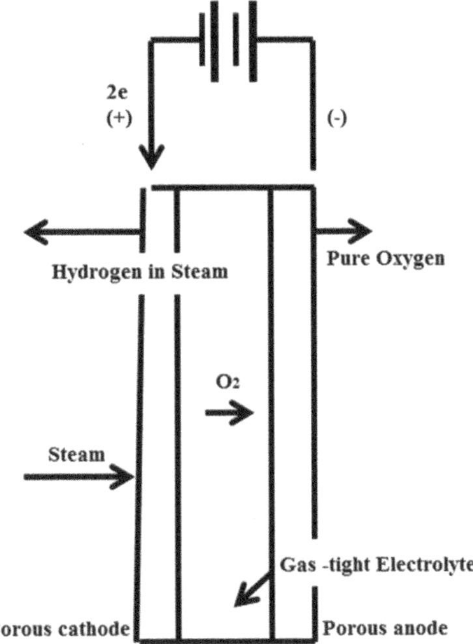

FIGURE 2.4 High-temperature electrolysis scheme.

cathode, where they combine with free electrons to produce heat and water [22–25]. These membranes are extensively utilized in fuel cells to generate electricity and in electrolyzers for making hydrogen. The PEMs act as a way of separating the cathode from the anode. Generally, a PEM is used for the conduction of protons from the anode to the cathode. The most commonly used membranes for this purpose are Nafion™ and Nafion™-based due to their thermostability, high ionic conductivity, excellent chemical stability, good mechanical strength, and durability at low temperatures under high levels of relative humidity [26–29]. Nafion™ and Nafion™ membranes have drawbacks of poor conductivity at low humidity [30,31] and a drop in ionic conductivity at 90°C under low relative humidity [32–34], respectively. The advantages of PEM electrolyzers are their capability to work at high current densities with high voltage and to yield pure hydrogen gas of 99.995% purity [35]. The problems with using a PEM electrolysis system are the elevated cost of catalysts and the necessity for a costly membrane that has only normal stability. Also, PEM electrolyzer stack materials are more expensive than alkaline ones [36,37]. A schematic representation of the PEM electrolysis route is revealed in Figure 2.5.

2.5 PHOTO-ELECTROCHEMICAL

The photo-electrochemical (PEC) method for hydrogen production is a novel and talented approach that utilizes the principles of electrochemistry, photochemistry, and materials science to generate hydrogen from water using sunlight as an energy

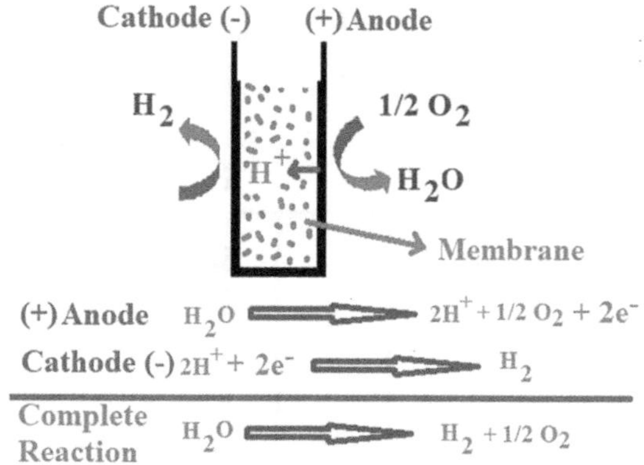

FIGURE 2.5   Proton exchange membrane (PEM) electrolysis.

source. This method usually includes the utilization of a specific device called a photo-electrochemical cell or PEC cell [38].

A photo-electrode, frequently made of semiconducting materials, is exposed to sunlight. The photo-electrode absorbs photons from sunlight, making electron-hole pairs (excitons) within the material. The electron-hole pairs made by photon absorption are separated within the photo-electrode. The electrons pass to the conduction band, while holes move to the valence band. The electrons are free to conduct electricity [39]. The electrons created in the photo-electrode are utilized to drive the electrochemical reduction of water at the photo-electrode surface. The water-splitting reaction includes the reduction of water at the cathode, which generates hydrogen gas (H_2), and the oxidation of water at the anode, which generates oxygen gas (O_2).

$$\text{Cathode (Reduction)}: 2H_2O + 2e^- \rightarrow 2OH^- + H_2 \tag{2.3}$$

$$\text{Anode (Oxidation)}: 2H_2O \rightarrow O_2 + 4H^+ + 4e^- \tag{2.4}$$

An electrolyte is used to assist the transport of protons (H^+) between the anode and cathode, permitting the water-splitting reactions to happen. The selection of electrolytes depends on the precise PEC cell design. Hydrogen gas generated at the cathode is collected for use as a clean and renewable energy source [40].

The advantages of this method are given below:

- Utilizes sunlight, making it environmentally friendly.
- It can be used for on-site and on-demand hydrogen manufacture.
- PEC cells can be designed to be highly effective.

A schematic representation of photo-electrochemical hydrogen production is shown in Figure 2.6.

The major challenges and limitations are given below:

- The development of effective and stable photo-electrode materials is a main challenge.

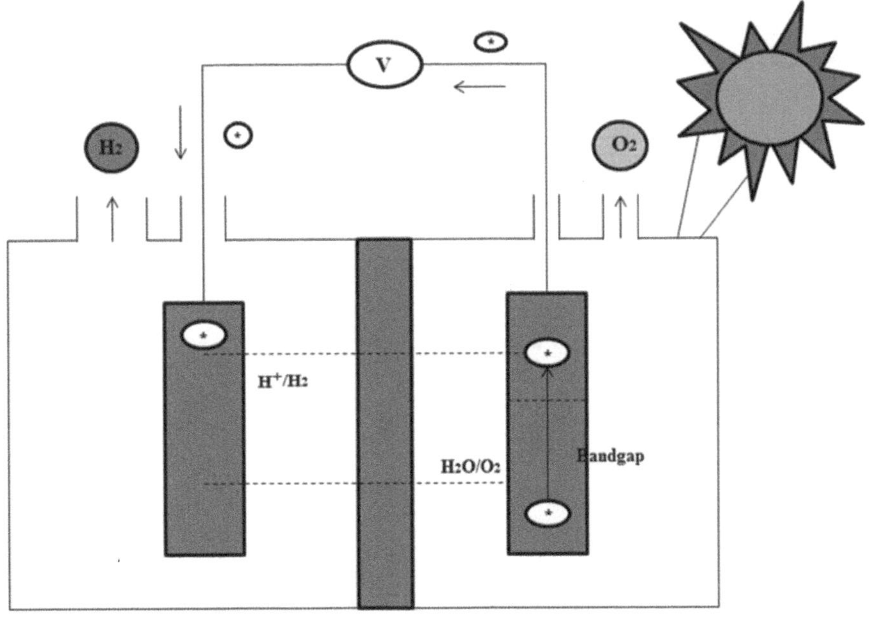

FIGURE 2.6   A schematic representation of photo-electrochemical hydrogen production.

- PEC cells are difficult to design and operate.
- PEC systems may need specific constituents and materials, making them expensive to implement.

The research into PEC technology is continuing, with a focus on refining materials, increasing efficiency, and dropping costs. PEC is measured as a promising method for sustainable and green hydrogen manufacture, principally in applications where sunlight is ample, such as solar hydrogen production.

2.6 PHOTOCATALYTIC (PC)

Hydrogen production by water splitting in the presence of suitable radiation and catalysis is the most important method. It has gained importance in hydrogen production in recent years. This is due to the fact that water is freely available in nature. Also, the sun's radiation may be used to split water into hydrogen and oxygen. It is important to mention that sun radiation is also freely available. The formula of water is H_2O and it is clear that it is made of hydrogen and oxygen atoms, which are split into hydrogen and oxygen atoms. The atoms recombine individually and respectively resulting in hydrogen and oxygen gases. These gases are separated and used for further beneficial purposes. The chemical equation of water splitting is shown in Figure 2.7.

The first paper on water radiolysis was published by Debierne [41]. The author hypothesized that the radiolysis of water proceeded through the production of the H● atom and the ●OH radicals, which may later on give hydrogen gas. Later on, in 1973, Fujishima and Honda used TiO_2 electrodes and published the first paper on water

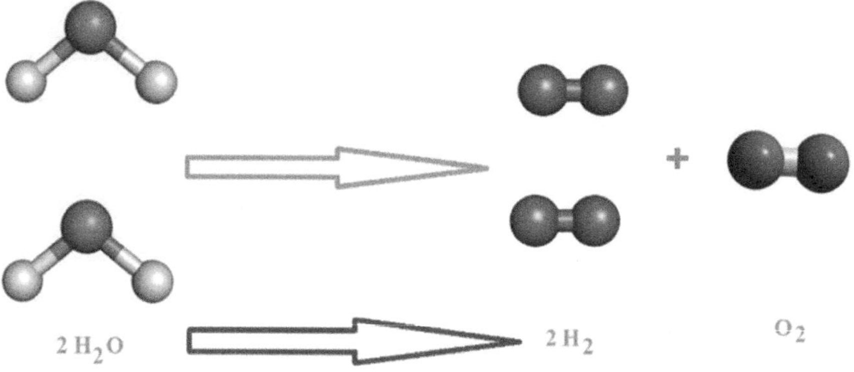

FIGURE 2.7   Chemical equation of water splitting.

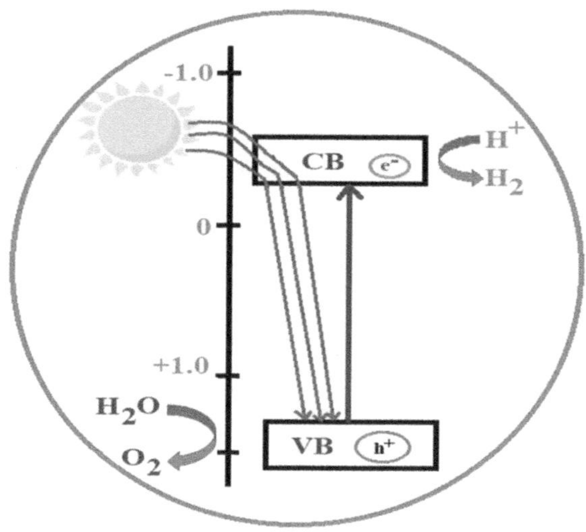

FIGURE 2.8   A schematic representation of radiation decomposition of water.

decomposition using UV radiation [42]. Water splitting is an essential reaction in many technological applications, for example in fuel cells, solar energy production, and catalysis. This process can be more efficient if it is assisted by nano photocatalysts suspended directly in water. The schematic representation of radiation decomposition of water is shown in Figure 2.8.

2.7   RADIOLYSIS

The concept of water radiolysis was coined from the work of Debierne who first time hypothesized that water radiolysis proceeded by the production of H$^•$ and $^•$OH radicals [41]. Water radiolysis states water decomposition by utilizing the radiation. The most commonly used radiations are gamma, alpha, X-rays, and other high-energy particles. This decomposition leads to the water splitting into hydrogen and oxygen. The hydrogen production in radiolysis is shown below.

$$H_2O + \text{Ionising radiation} \rightarrow e_{aq}^- + H^\bullet + HO^\bullet + H_3O^+ + HO_2^\bullet + H_2O_2 + OH^- + H_2 \quad (2.5)$$

The hydrogen production in radiolysis includes numerous steps as given below. These steps are also considered as the tentative mechanism of hydrogen generation in radiolysis [43].

$$H_2O + e^- \rightarrow e_{aq}^- \quad (2.6)$$

$$e_{aq}^- + H^+ \rightarrow H^\bullet \quad (2.7)$$

$$e_{aq}^- + e_{aq}^- + 2H_2O \rightarrow H_2 + 2OH^- \quad (2.8)$$

$$e_{aq}^- + H^\bullet + H_2O \rightarrow H_2 + OH^- \quad (2.9)$$

$$H^\bullet + H_2O \rightarrow H_2 + HO^\bullet \quad (2.10)$$

$$H^\bullet + H^\bullet \rightarrow H_2 \quad (2.11)$$

As evident from the above mechanism, water radiolysis can also yield other chemical species, such as eaq^-, $H^\bullet$, $HO^\bullet$, H_3O^+, $HO_2^\bullet$, H_2O_2, OH^-, and H_2 [44]. Hydrogen generation by water radiolysis has been studied as a possible way of producing hydrogen clean and efficient fuel. Nevertheless, the practical presentations of this method face challenges and issues, including the necessity for high-energy sources and the separation of hydrogen from other molecules and radicals.

Numerous factors affect the efficacy and practicality of hydrogen manufacture through water radiolysis, like the types of radiation utilized, the materials involved, and the total energy competence of the procedure. The research in this field remains to explore the ways to advance the competence and probability of utilizing water radiolysis for hydrogen manufacture as part of the broader effort to develop sustainable and clean energy [45].

REFERENCES

1. P. Sean, NASA's Perseverance mars rover extracts first oxygen from red planet, *NASA*, 2021–04–22 (2021).
2. K. Liu, C. Song, V. Subramani, *Hydrogen and Syngas Production and Purification Technologies*, John Wiley & Sons, Inc. (2009).
3. S. Farid, D. Ibrahim, A review and comparative evaluation of thermochemical water splitting cycles for hydrogen production, *Energy Convers. Manag.*, 205, 112182, (2020).
4. G.W. Crabtree, M.S. Dresselhaus, M.V. Buchanan, The hydrogen economy, *Phys. Today*, 57, 39–44 (2004).
5. W.W. Akers, D.P. Camp, Kinetics of the methane-steam reaction, *AIChE J.*, 1(4), 471–475 (1955).
6. J. Xu, Methane steam reforming, methanation and water-gas shift: I. Intrinsic kinetics, *AIChE J.*, 35(1), 88–96 (1989).

7. G. Franchi, M. Capocelli, M.D. Falco, V. Piemonte, D. Barba, Hydrogen production via steam reforming: A critical analysis of MR and RMM technologies, *Membranes*, 10, 10 (2020).

8. M. Mosinska, M.I. Szynkowska-Józwik, P. Mierczynski, Catalysts for hydrogen generation via oxy–steam reforming of methanol process, *Materials*, 13, 5601 (2020).

9. S.S.S.A. Khusaibi, L.N. Rao, Design and production of hydrogen gas by steam methane reforming process - A theoretical approach, *Int. J. Sci. Technol. & Eng.*, 3, 472–476 (2016).

10. B. Lu, K. Kawamoto, Thermo-chemical hydrogen production technology from biomass, *Curr. Org. Chem.*, 19, 447–454 (2015).

11. G. Ji, J.G. Yao, P.T. Clough, J.C.D.D. Costa, E.J. Anthony, P.S. Fennell, W. Wang, M. Zhao, Enhanced hydrogen production from thermochemical processes, *Energy Environ. Sci.*, 11, 2647–2672 (2018).

12. A. Boretti, J. Nayfeh, A. Al-Maaitah, Hydrogen production by solar thermochemical water-splitting cycle via a beam down concentrator, *Front. Energy Res.*, 9, 666191 (2021).

13. R.D. Levie, The electrolysis of water, *J. Electroanal. Chem.*, 476, 92–93 (1999).

14. D. John, On some chemical agencies of electricity, *Collec. Works Sir Hum. Dav.*, 5, 1–12, (1839).

15. L.D. Aleksandrovich, 26 July 2011, Encyclopedia. https://web.archive.org/web/20110726201906

16. Z. Steven, *Chemistry* (9th ed.). Cengage Learning (1 January 2013). p. 30. ISBN 978-1-13–361109-7.

17. M. Carmo, D.L. Fritz, J. Mergel, D. Stolten, A comprehensive review on PEM water electrolysis, *Int. J. Hydrogen Energy*, 38(12), 4901–4934 (2013).

18. Hydrogen Basics — Production. Retrieved 5 February 2008, http://www.sciepub.com/reference/40772

19. A. Hauch, S.D. Ebbesen, S.H. Jensena, M. Mogensena, Highly efficient high temperature electrolysis, *J. Mater. Chem.*, 18(20), 2331–2340 (2008).

20. S.P.S. Badwal, S. Giddey, C. Munnings, Hydrogen production via solid electrolytic routes, *WIREs Energy Environ.*, 2(5), 473–487 (2012).

21. K. Yamada, S. Makino, K. Ono, K. Matsunaga, M. Yoshino, T. Ogawa, S. Kasai, S. Fujiwara, High temperature electrolysis for hydrogen production using solid oxide electrolyte tubular cells assembly unit, *AICHE Annual Meeting* (2006).

22. K. Hou, R. Hughes, The kinetics of methane steam reforming over a Ni/α-Al2O catalyst, *Chem. Eng. J.*, 82(1), 311–328 (2001).

23. L.J. Nuttall, A.P. Fickett, W.A. Titterington, In: Hydrogen energy; Proceedings of the Hydrogen Economy Miami Energy Conference, Miami Beach, Fla., March 18-20, 1974. Part A. (A75-44751 22-44) New York, Plenum Press, 1975, p. 441–455 (1975).

24. R.L. LeRoy, M.B.I. Janjua, R. Renaud, U. Leuenberger, Analysis of time-variation effects in water electrolyzers, *J. Electrochem. Soc.*, 126(10), (1679).

25. A.H.A. Rahim, A.S. Tijani, S.K. Kamarudin, An overview of polymer electrolyte membrane electrolyzer for hydrogen production: Modeling and mass transport, *J. Power Sources*, 309, 56–65 (2016).

26. A. Rahim, An overview of polymer electrolyte membrane electrolyzer for hydrogen production: Modeling and mass transport, *J. Power Sources*, 309, 56–65 (2016).

27. J. Escorihuela, A. García-Bernabé, V. Compañ, A deep insight into different acidic additives as doping agents for enhancing proton conductivity on polybenzimidazole membranes, *Polymers*, 12, 1374 (2020).

28. J. Escorihuela, A. García-Bernabé, Á. Montero, A. Andrio, Ó. Sahuquillo, E. Giménez, V. Compañ, Proton conductivity through polybenzimidazole composite membranes containing silica nanofiber mats, *Polymers*, 11, 1182 (2019).

29. X. Sun, K. Xu, C. Fleischer, X. Liu, M. Grandcolas, R. Strandbakke, T.S. Bjørheim, T. Norby, Earth-abundant electrocatalysts in proton exchange membrane electrolyzers, *Catalysts*, 8, 657 (2018).
30. Y.P. Ying, S.K. Kamarudin, M.S. Masdar, Silica-related membranes in fuel cell applications: An overview, *Int. J. Hydrogen Energy*, 43, 16068–16084 (2018).
31. D.J. Kim, D.H. Choi, C.H. Park, S.Y. Nam, Characterization of the sulfonated PEEK/sulfonated nanoparticles composite membrane for the fuel cell application, *Int. J. Hydrogen Energy*, 41, 5793–5802 (2016).
32. J. Escorihuela, Ó. Sahuquillo, A García-Bernabé, Phosphoric acid doped polybenzimidazole (PBI)/Zeolitic imidazolate framework composite membranes with significantly enhanced proton conductivity under low humidity conditions, *Nanomaterials*, 8, 775 (2018).
33. X. Sun, S.C. Simonsen, T. Norby, A. Chatzitakis, Composite membranes for high temperature PEM fuel cells and electrolysers: A critical review, *Membranes*, 9, 83 (2019).
34. Y. Dong, J. Liu, M. Sui, Y. Qu, J.J. Ambuchi, H. Wang, A combined microbial desalination cell and electrodialysis system for copper-containing wastewater treatment and high-salinity-water desalination, *J. Hazard. Mater.*, 321, 307–315 (2017).
35. C.N.C. Hitam, A.A. Jalil, A review on biohydrogen production through photofermentation of lignocellulosic biomass, *Biomass Convers. Biorefin.*, 13, 8465–8483 (2023).
36. M.F.A. Kamaroddin, N. Sabli, P.M. Nia, T.A.T. Abdullah, L.C. Abdullah, S. Izhar, A. Ripin, A. Ahmad, Phosphoric acid doped composite proton exchange membrane for hydrogen production in medium-temperature copper chloride electrolysis, *Int. J. Hydrogen Energy*, 45, 22209–22222 (2020).
37. M. Luo, Y. Yi, S. Wang, Z. Wang, M. Du, J. Pan, Q. Wang, Review of hydrogen production using chemical-looping technology, *Renew. Sustain. Energy Rev.*, 81, 3186–3214 (2018).
38. P. V. Kamat, K. Sivula, Photoelectrochemical and photocatalytic hydrogen generation: A virtual issue, *ACS Energy Lett.*, 7(12), 4379–4380 (2022).
39. B.D. Sherman, N.K. McMillan, D. Willinger, Sustainable hydrogen production from water using tandem dye-sensitized photoelectrochemical cells, *Nano Convergence* 8, 7 (2021).
40. N. Iqbal, M.S. Khan, M. Zubair, S.A. Khan, A. Ali, N. Aldhafeeri, S. Alsahli, M. Alanzi, A. Enazi, T. Alroyle, A. Alrashidi, Advanced photoelectrochemical hydrogen generation by CdO-g-C3N4 in aqueous medium under visible light, *Molecules*, 27, 8646 (2022).
41. A. Debierne, Recherches sur les gaz produits par les substances radioactives, *Ann. Phys.*, 2, 97–127 (1914).
42. A. Fujishima, Electrochemical photolysis of water at a semi-conductor electrode, *Nature*, 238, 37–38 (1972).
43. W. Wang, C. Liu, W. Liu, D. Zhang, Factors influencing hydrogen yield in water radiolysis and implications for hydrocarbon generation: A review, *Arabian J. Geosci.*, 11, 542 (2018).
44. M.E. Dzaugis, A.J. Spivack, A.G. Dunlea, R.W. Murray, S. D'Hondt, Radiolytic hydrogen production in the subseafloor basaltic aquifer, *Front. Microbiol.*, 2016, 76 (2016).
45. S.L. Caër, Water radiolysis: Influence of oxide surfaces on H2 production under ionizing radiation, *Water*, 3, 235–253 (2011).

3 Advanced nanomaterials for water splitting

3.1 INTRODUCTION

Water splitting is a critical process for the production of clean and sustainable hydrogen energy. The water splitting is achieved by electro-catalysts, photocatalysts, and photo-electrochemical (PEC) methods. All three methods need a catalyst to provide a good amount of hydrogen at a large scale inexpensively. Nowadays, nanomaterials are being used as catalysts and in the case of photocatalysts, and PEC methods, and these are called photocatalysts. These photocatalysts play a significant role in water splitting as essential materials for water splitting. Many materials have been prepared and used for water splitting. The most important are nanoparticles (NPs), quantum dots (QDs), perovskite materials, 2D materials, doped materials, hybrid materials, molecularly engineered materials, etc. These advanced nanomaterials have the potential to significantly enhance the efficiency and cost-effectiveness of water-splitting technologies, which are essential for the production of clean hydrogen as a sustainable energy carrier. Ongoing research and development in this field aim to bring these advanced nanomaterials into practical applications and contribute to a greener and more sustainable energy future. Some of the materials used in water splitting are discussed in this chapter. Besides, the effect size, surface area, and surface charges are discussed.

3.2 DIFFERENT NANOMATERIALS PREPARED BY CHEMICAL ROUTES

Nanomaterials can be prepared by various chemical routes. These methods are essential for controlling the size, shape, composition, and properties of the nanomaterials. The most important methods are chemical precipitation, sol-gel synthesis, chemical vapor deposition, hydrothermal synthesis, microwave-assisted synthesis, chemical reduction, emulsion techniques, template-assisted synthesis, self-assembly, and electrochemical synthesis. The choice of the method depends on the desired material, its properties, and the specific applications. Researchers often combine various techniques to achieve the desired nanomaterial characteristics and functionalities. Some of the materials prepared by chemical routes and used in water splitting are discussed in the following paragraphs.

Nanomaterials have received a lot of interest in recent years due to their different chemical and physical properties compared to bulk materials. They also have various applications in different technologies including energy [1]. Nanomaterial-based photo-electrodes have been explored for various types of applications. Often using

DOI: 10.1201/9781003432364-4

single-component nanomaterials like metal chalcogenides are frequently preferred materials due to their exceptional hydrogen-producing catalytic activity and acceptable band gaps. Sometimes, nanomaterials can increase specific surface areas and reactive sites, which improves surface catalysis. Nonetheless, particle size decreases until the quantum confinement effect occurs, which increases the e/h pair recombination ratio. As a result, the migration of e/h pairs needs a decreased recombination of enough potential gradient from the core to the surface of the materials, which is linked to their surface and structural properties [2]. Metal oxides have also been widely explored as photocatalysts due to their excellent combination of electronic structure, light absorption qualities, charge transport characteristics, and excited lifetimes. The important metal oxide nanomaterials for this purpose are TiO_2 [2–18], In_2O_3 [19], Sb_2S_3 [20] Bi_2S_3 [21], Fe_2O_3 [22,23], ZnO [24,25], WO_3 [26–31], CdS [32,33], CuO, Bi_2O_3, and MoO_3. Among them, TiO_2 and ZnO nanomaterials are the most commonly used materials for water spitting and hydrogen production. These nanomaterials have a bandgap of 3.03–3.37 eV resulting in low efficiency in solar light absorption, but their stabilities and low price make them potential candidates in water slitting. They are chemically stable, very transparent and have high electron mobilities. The formation of hydrogen is affected by the shape, defect density, and electrolyte interaction [34–45].

In addition, the coupling of 1D metal oxide nanostructures with 0D narrow band gap semiconductor photosensitizers such as CdS and CdSe NPs or QDs has recently received a lot of interest [32,33]. QDs including CdS [46,47], CdSe, CdTe [48], PbS, Ag_2S, and $CuInS_2$ have been shown to increase the photo-activity of catalysts. QDs have a narrow band gap and can be used as dye sensitizers in the visible spectrum to sensitize semiconductors with a broad band-gap. Under solar light, CdS/vermiculite has a much greater water-splitting efficiency [quantum efficiency (QE): 17.7%] than CdS/attapulgite (QE: 8.2%). Cadmium sulfide with tiny band gap energy (2.4 eV) may change the broad band gap of vermiculite (3.5 eV). Its hydrogen production rate can reach 92 mmol/h, which is ten times that of pure CdS. The photocurrent density of CdS/vermiculite is 0.4 mA/cm^2 at 1.90 V, which is five times that of bare vermiculite [47]. Some of the nanomaterials used for hydrogen production are listed in Table 3.1.

Moniruddin et al. [63] reported that strontium titanate ($SrTiO_3$) is one of the most promising materials as a photocatalyst to produce hydrogen gas from water splitting. The authors used an electrospinning technique and the sol-gel method to develop 3D porous $SrTiO_3$ nanostructures. $SrTiO_3$-nanofibers (STO-NFs) of varied crystallite sizes were synthesized by modifying the synthesis parameters, such as the precursor concentration and calcination temperature. The amount of H_2 generated by water splitting under UV irradiation was used to evaluate the photocatalytic activity of different STO-NFs (depending on crystallite size). A hydrogen evolution study revealed that the photocatalytic activity of STO-NFs was significantly dependent on crystallite size, precursor concentration, and calcination temperature. The photocatalytic activity of STO-NFs was compared to that of commercial $SrTiO_3$ NPs (STO-NPs) after Pt was added as a co-catalyst. The NFs revealed a two times higher H_2 production (1.14 mmol/g.h) over the NPs. Akyüz et al. [64] synthesized SWCNT hybrid nanomaterial [Single walled carbon nanotubes (SWCNT)-Coum-Pc] and used it for

TABLE 3.1

Hydrogen evolution by nanomaterials

Nanomaterials	Hydrogen production rate	Ref.
CdS	0.31 mmol/h/g	[49]
MoS_2eCdS	5.24 mmol/h/g	[50]
1D CdS nanowire/2D MoS_2 nanosheet	9.73 mmol/h/g	[51]
CdS/ Graphene quantum dots (GQDs) nanohybrids	95.4 mmol/h	[52]
CD-CdS	2.55 mmol/h	[53]
Nanostructured CdS	2,945 mmol/h	[54]
CdS QD-sensitized TiO2/Pt	0.8 mmol/h/g	[47]
Pure TiO_2	18.74 mmol/h/g	[55]
GQDs/TiO_2	41.26 mmol/h/g	[55]
TiO_2 Nanoparticle	61.85 mL/h/cm^2	[56]
TiO_2 thin film	59.8 mmol after 8 h	[57]
CD-TiO_2 nanoparticle	110.45 mL/h/cm^2	[56]
TiO_2eNi$(OH)_2$/CdS	5.36 mmol/h/g	[58]
TiO_2Ni$(OH)_2$/CNT/CdS	12 mmol/h/g	[58]
CNTs/C_3N_4	44.64 mmol/h/g	[59]
CNTs/$MnO_2C_3N_4$	122 mmol/h/g	[59]
Pure MoS_2 spheres	881.6 mmol/h/g	[60]
Au multimer@MoS_2 spheres	2,997.2 mmol/h/g	[60]
Mesoporous $SrTiO_3$ nanocrystal	188 mmol/h/g	[61]
SiC nanoparticle	0.85 mmol/g/h	[62]
$SrTiO_3$ nanocrystal	276 mmol/h/g	[61]

water splitting. In comparison to the SWCNT catalyst, the SWCNT-Coum-Pc hybrid decreased the onset potential required for HER by around 200 mV. SWCNT-Coum-Pc hybrid considerably enhanced the water-splitting abilities of OER and HER. Fazil et al. [65] reported the synthesis of pristine-TiO_2 and Sr-doped TiO_2 (1%, 2.5%, and 5%) nanomaterials via the hydrothermal method for water splitting. Sr-doped TiO_2 showed good performance in water splitting. A high photo-oxidation intensity and distinctive electronic structures with a relatively small band gap are found in bismuth-based nanomaterials, which also showed exceptional stability in an aqueous medium. These features provide a rich source of electrons and photons that can be very helpful for the process of producing hydrogen by water splitting (Figure 3.1).

Li et al. [66] reported the synthesis of bismuth oxychloride nanomaterial by using potassium chloride and bismuth nitrate as the main precursors through a hydrothermal method assisted by sonication. The authors reported a good water-splitting performance. Ye et al. [67] synthesized bismuth oxy chloride ultrathin nano-sheets by using the solvothermal method for water-splitting purposes. Similarly, Zhang et al. [68] synthesized NiOX-doped BiOCl ultrathin nanomaterial (3.6 nm size) for hydrogen production. The best performance was achieved by BiOCl nanomaterial.

Khan et al. [69] reported the synthesis of multi-ferroic terbium ortho-ferrite NPs at low temperature via polymeric citrate precursor route. The authors reported

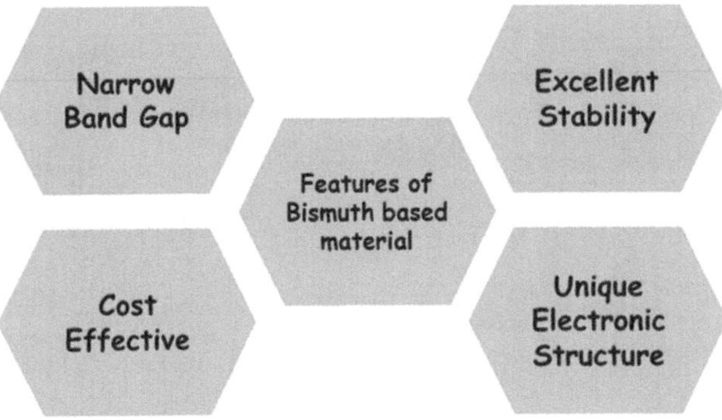

FIGURE 3.1 Salient features of bismuth-based materials for water splitting.

1.44 mmol/h.g by using these NPs. This group is very active in this area and produced various papers including synergistic effect of multi-ferroicity in $GdFeO_3$ NPs for significant hydrogen [70] and bifunctional multiferroic $GdCrO_3$ nano-assemblies for sustainable H_2 production [71]. Furthermore, the same workers prepared hydrothermally derived Mg-Doped TiO_2 nanostructures and used them for augmented hydrogen production [72]. Besides, pristine and Ag-doped WO_3 nanoplates were prepared and used for the same purpose [73]. Additionally, the symbiotic MoO_3-$SrTiO_3$ hetero-structured nano-catalysts were prepared and used for sustainable hydrogen production [74]. Besides, other publications are also available from this group in this research area.

3.3 DIFFERENT NANOMATERIALS PREPARED BY GREEN ROUTES

The technology for preparing nanomaterials by green routes is called green nanotechnology. This involves the synthesis and application of nanomaterials using environmentally friendly and sustainable methods. These methods aim to minimize the use of hazardous chemicals, reduce waste, and lower energy consumption. Green nanotechnology is crucial for addressing environmental and health concerns associated with conventional nanomaterial synthesis and applications. Some of the important methods of green nanomaterials involve biological (bio-fabrication), microwave, ultrasound-assisted, supercritical fluid technology, green NP coating, green nanocomposites, recycled nanomaterials, sustainable sourcing, etc. methods.

Green nanomaterials are known as the materials that minimize environmental concerns while simultaneously improving human health. Green nanotechnology develops NPs with enhanced properties for a variety of uses. The main aim of producing new NPs is to improve sustainability and make them more environmentally friendly. A wide range of NPs may be synthesized by combining several biological identities such as algae, bacteria, yeast, fungi, plants, and viruses. Most nanomaterials are made by oxidizing and reducing metallic ions with enzymes, alkaloids, polyphenols, terpenoids, sugars, proteins, aldehydes, carboxyl groups, and so on. Recently, *in vitro* techniques for the bio-reduction of metal ions into nanomaterials have been

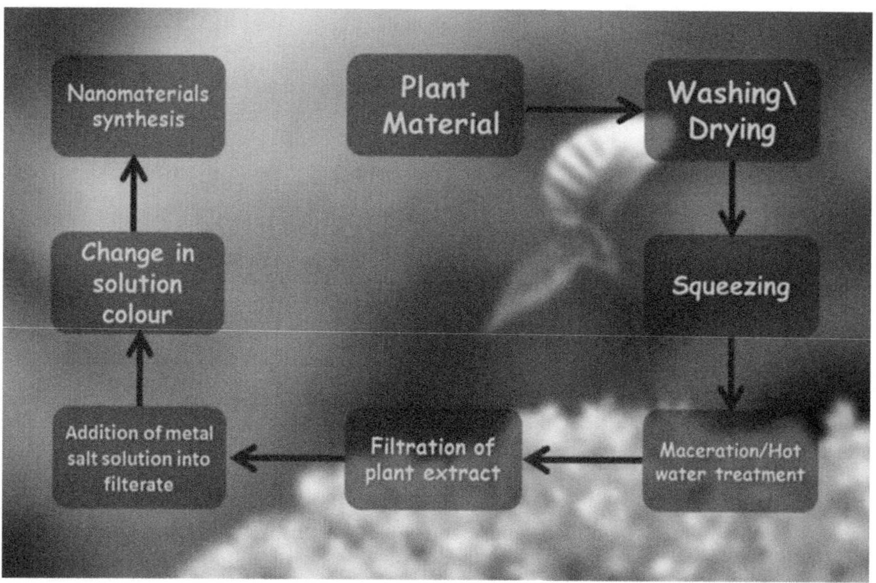

FIGURE 3.2   The schematic representation of nanomaterial formation via green technology.

developed by using plant extracts. These approaches are not only eco-friendly and affordable, but also experimentally helpful for controlling the nanomaterial's size and shape. Nanomaterials of various metal ions have been synthesized by using various kinds of plants [75–78]. The most significant metal ions utilized to create diverse nanomaterials prepared by the green approach include titanium, zinc, iron, copper, silver, cobalt, palladium, gold, antimony, selenium, and so on [79].

The plant components such as leaves, stems, fruits, roots, calluses, seeds, flowers, and peels are used to make nanomaterials. Nowadays, the synthesis of nanomaterials using microorganisms through bio-mineralization or biological processes is gaining popularity. For the synthesis of nanomaterials, the biological processes of controlled mineralization and induced mineralization are well recognized [80–83]. Under anaerobic or aerobic conditions, the interactions of organic matter (electron donor) with metal ions (electron acceptor) are primarily responsible for the creation of nanomaterials [84–86]. The schematic representation of nanomaterial formation via green technology is shown in Figure 3.2.

Ismail et al. [87] synthesized Ru(IV) oxide NPs with a 2.15 nm average diameter by using *Aspalathus linearis* natural extract. The material was found to have a 2.1 eV band. RuO_2-Cu_2O nanomaterials were prepared by loading RuO_2 on Cu_2O thin film. These showed effective photo water splitting under the solar spectrum. Realpe et al. [88] reported the bio-synthesis of TiO_2 with lemon grass extract. The band gaps ranged from 3.08 to 2.66 eV. Raorane et al. [89] synthesized copper-doped TiO_2 NPs using *Annona squamosal* fruit peel extract. The band gap was between 2.17 and 2.50 eV, shifting the absorption to the visible range. The authors reported enhanced photocatalytic activity, which might be utilized for producing hydrogen. Alajmi et al. [90] reported the synthesis of Fe_3O_4 nanomaterials with a surface area of 150 m^2/g, by using the extract of *Pandanus odoratissimus* leaves. The electrochemical water splitting worked effectively with a good yield. Selvanathan et al. [91] reported the synthesis of nickel oxide NPs (NiO_x NPs) using phytochemicals from three separate sources, i.e., aloe vera leaf, papaya peel, and dragon fruit peel extracts. It was observed that NPs obtained by aloe vera-mediated NiOx have large surface area, high conductivity, and effective charge-transfer kinetics, which contributed to their low over-potential values. Zahra et al. [92] reported the biosynthesis of stable NiO using *Olea ferruginea* as a reducing agent. The NiO nanomaterial has an average particle size of 13 nm with a uniform cubic shape. NiO with 0.41 V potential was found to be comparable to benchmark catalysts. Zhang et al. [93] prepared $g\text{-}C_3N_4\text{-}Pt$ photocatalyst using a biosynthetic method. The material used was to evolve hydrogen by water splitting. Zahra et al. [94] reported biosynthesis of ZnO, ZrO_2, and PdO NPs by using *Olea ferruginea royle* foliar extract. The average crystallite sizes of ZnO, ZrO_2, and PdO were 19.83, 11.17, and 16.5 nm, respectively.

3.4 GRAPHENE AND ITS DERIVATIVES

It is witnessed that plastic was the material of the 20th century but graphene is the material of this century. Graphene was developed by Andre Geim and Konstantin Novoselov for which these researchers got a Nobel Prize in 2010. Since its discovery, graphene has been gaining importance and attention continuously in various research areas, i.e., catalyst, energy, sorbent, sensors, electrics, electronics, textile, robotics, etc. Graphene has a wide range of applications because of its special features [95–98], which are summarized below:

- Two hundred times stronger than steel with light-weight.
- Immensely tough and flexible and can be bent or folded.
- Perfect barrier due to its compact structure. Not even helium can pass through it.
- Transparent due to its tiny thickness.
- Absorbs only 2.3% of light and the rest is transmitted through and it is almost invisible.
- The thinnest material that is known to exist (surface area = ~2,700 m^2/g).
- Electrically and thermally conductive with better conductivity than copper (1,000 times copper).
- Since carbon is the Earth's fourth most common abundant element graphene is very inexpensive.
- Stretchable changing in sheet and stretched by as much as 20% without inducing defects.
- A stretching sheet can alter its magnetic properties.
- Chemically inert and graphene derivatives have a wide range of physical and chemical properties.

After analyzing the above-mentioned graphene features, it can be estimated that graphene may be used in almost all areas of research and applications including water splitting for hydrogen production. Of course, graphene has a wide range of applications but we narrowed down this chapter to hydrogen production by water splitting because of the great demand to develop inexpensive green energy hydrogen [99–102]. Graphene sheets can accept and transport electrons from excited semiconductors, suppress charge recombination, improve interfacial charge transfer processes, and provide significantly more active adsorption sites and photocatalytic reaction centers, resulting in increased photocatalytic H_2 production activity.

Because of its large surface area and fast electron mobility, graphene can serve as a suitable mediator and sink for electrons, as well as a surface for homogenous dispersion of other semiconductors. However, the gap between the valence and conduction bands is zero [103], since antibonding p* orbitals and p orbitals meet at the Brillouin zone corners. The nanocomposites or other derivatives of graphene are required to get suitable band gaps. The band gaps may be accomplished by adjusting the structure by introducing heteroatoms or functional groups [104]. Doping with heteroatoms such as nitrogen, sulfur, phosphorus, and others causes a change in the electronic density of the graphene structure.

Graphene can be prepared by bottom-up or top-down [105,106]. Graphene synthesized via liquid exfoliation technique [107] has a condensed surface area due to the p-p interactions that occur between the layers. Graphene has been studied in several forms, including nanoribbons or nano graphite ribbons with ultra-thin widths of less than 2–5 nm [108]. The zigzag graphene nanoribbons may be synthesized by unzipping multi-walled carbon nanotubes with $KMnO_4$ and H_2SO_4 [109]. The combination of graphene and TiO_2 resulted in a lower recombination rate, which improves the photocatalytic efficacy [110]. The development of a hybrid structure with graphene decreases the recombination rate, narrows the band gap, and increases oxygen vacancies, all of which improve the rate of H_2 generation. The various photocatalysts have variable tendencies to reduce the recombination rate of graphene's charge carriers, hence the rate of hydrogen generation by different photocatalysts with graphene also differs. Graphene sheet aggregation was also prevented by hybridizing it with CdS. The synthesis of graphene oxide (GO) is caused by the surface of graphite being embellished with oxygen-containing groups. This task was achieved by sonicating exfoliation from layered sheets of the graphite oxide structure [111].

The use of graphene-based hybrid nanocomposites in H_2 evaluation methods is described [112]. Photoexcited electrons enter the graphene's LUMO and move to the surface through π^* orbitals, resulting in the greatest possible charge separation. Carbon materials are good electron acceptors and transit routes for limiting photo-induced electron-hole pairs and increasing the rate of photocatalytic H_2 reaction [113]. Due to a high charge separation between VB and CB, metal-doped graphene materials have remarkable photocatalytic properties. The nitrogen-doped graphenes are substantially superior to those of pure graphene and, hence, gaining greater attention [114]. Interest is growing in the addition of TiO_2 photocatalysts onto graphene. It allows easy-of-charge transit separation, and a wider light absorption range [115]. When exposed to UV-visible light a G/TiO_2 composite photocatalyst with 5% graphene produced hydrogen (86 mmol/h/g) [116].

Nitrogen-doped graphene is used to produce hydrogen at 6 mmol/h. This was due to the fact that a nitrogen-doped semiconductor contains both n- and p-type conductivities and, consequently, the photocatalytic efficiency is improved [117,118].

Cu_2O has a 2.2 eV band gap and is inexpensive, making it a viable photocatalyst. However, Cu_2O's photo-generated electrons and holes are easily recombined, which is an important issue for the progression of redox reactions. The hybrid graphene/ Cu_2O material showed outstanding semiconducting qualities and effective hydrogen evaluation [119]. Also, Cu_2O/doped graphene composite generated hydrogen at a rate of 19.5 mmol/h under visible light [120]. Park et al. [121] synthesized Pt-TiO_2, GO-TiO_2, and a ternary hybrid of GO with Pt and TiO_2. The authors discovered that GO served as a co-catalyst and improved H_2 evolution. TiO_2 nano-sheets modified by GO were synthesized by Xiang et al. [122] and the resulting hybrid material showed increased photocatalytic H_2 production when used with methanol as a sacrificial agent. Surprisingly, the H_2 production rate was increased to 736 mmol/ h/g with 3.1% quantum efficiency in the absence of the Pt co-catalyst, demonstrating 41 times greater performance than individual TiO_2 nano-sheets. Based on the reduced graphene oxide's (rGO) ability to transport charges and the efficacy of platinum NPs as electron traps for photo-generated electrons, the hydrogen efficiency of binary and ternary hybrid composites (TiO_2/rGO, TiO_2/Pt, and TiO_2/rGO/Pt) was evaluated. Adding rGO to TiO_2 nearly doubles the hydrogen generation rate (1.95%) compared to bare TiO_2, while the TiO_2/Pt photocatalytic system performed the best, with a 15.26-fold increase in hydrogen production compared to TiO_2 [123].

TABLE 3.2

Hydrogen generation by functionalized graphene derivatives

Composite materials	Co-catalysts	Sacrificial agent	Irradiation sources	Rates of H_2 production	Refs.
$LaNiO_3$-rGO	None	Methanol	UV-visible	3.22 mmol/h/g	[129]
rGO/Pt-TiO_2	Pt	None	Solar light	1,075.6 mmol/h/g	[130]
Graphene/ CdS	Graphene	Na_2S and Na_2SO_3	Xenon lamp	70 μmol/h/g	[125]
Graphene/ Cu_2O	None	None	Visible light	19.5 mmol/h/g	[121]
rGO/ $Zn_{0.8}Cd_{0.2}S$	None	Na_2S and Na_2SO_3	Solar irradiation	1,824 μmol/h/g	[127]
G/MoS_2/TiO_2	MoS_2	Ethanol	Xenon arc lamp	165.3 μmol/h/g	[127]
G/Bi_2WO_6	None	Methanol	Visible light	159.2μmol/h/g	[126]
GO/$ZnCl_2$/ Na_2S	Graphene	None	Visible light	7.42 μmol/h/g	[128]
N/doped graphene	None	Methanol	(neodymium-doped yttrium aluminum garnet; Nd:Y3Al5O12) ND-YAG laser	6 mmol/h/g	[118]
rGO/NiS/ $Zn_{0.5}Cd_{0.5}S$	None	None	Solar irradiation	375.5 μmol/h/g	[131]
TiO_2/doped graphene	Graphene	0.1 M Na_2S and 0.04 M Na_2SO_3	Visible light	86 mmol/h/g	[117]

TABLE 3.3

Advantages and disadvantages of graphene as photocatalysts

	Advantages	Disadvantages
Graphene	• Low cost • High surface area • Good electronic properties with electrical conductivity due to electron dislocation and its semimetal nature • Exclusive thermal conductivity • High Young's modulus • Extraordinary optical qualities since it can absorb light at a wide range of frequencies and radiation is absorbed regardless of frequency	• Zero band gap
Graphene oxide	• Good optical properties like the refractive index of 1.957 with an extinction coefficient of 0.101, and exhibiting complex dielectric behavior over λ range of 250–1,650 nm • Tunable structure • Exclusive mechanical properties, e.g., high Young's modulus ranging from 380 to 470 GPa, and high intrinsic strength • High surface area • High Young's modulus	• Poorly dispersible • Restocking nature • Multilayer thick structure • Insulating nature

Cadmium disulfide graphene clusters increase the surface area. This photocatalyst produced 70 mol/h hydrogen using a xenon lamp as an energy source [124]. Bi_2WO_6 with graphene produced hydrogen at a rate of 159.2 mol/h under UV light [125]. By using both graphene and MoS_2, the combined photocatalytic process efficiency was increased. Graphene/MoS_2/TiO_2 generated hydrogen in aqueous methanol at a rate of 165.3 mol/h/g using a xenon arc lamp [126]. When compared to pure GO, the photocatalytic effectiveness of GO/$ZnCl_2$/Na_2S was eight times higher at a 7.42 mol/h rate of hydrogen under visible light [127]. Nagaraju et al. [128] synthesized TiO_2 NPs and TiO-RGO hybrid nanomaterials via the ionothermal method. The graphene sheet's surface contains fixed TiO_2 NPs. TiO_2-RGO hybrid nanomaterials produce more H_2 (3.0 mmol/g) than pure TiO_2 nanomaterials (2.4 mmol/g). The hydrogen generation by some functionalized graphene materials is listed in Table 3.2.

Briefly, graphene is a marvelous material but it has certain limitations. The most important limitation is the zero band gap. There is a great demand to create a band gap in graphene because it has remarkable features needed for water splitting. Some of the advantages and disadvantages of graphene derivatives are summarized in Table 3.3.

3.5 SIZES OF DIFFERENT NANOMATERIALS

Nanomaterials have different properties as compared to bulk materials. The basic electrical and optical features of nanomaterials are size-dependent. These provide a range of properties that can be used in PEC water splitting. In 2002 Maier et al. [132] compared Mn oxides as normal and nano-materials. The later material was found to have entirely different redox potentials and water oxidation activities [133]. In 2010, Jiao et al. [134] reported that nanometer-sized Mn oxide clusters, supported on a mesoporous silica scaffold, were effective water oxidation catalysts in the aqueous solution at room temperature and pH 5. They also reported that the high-surface-area silica support may be important for the stability of the catalytic system because it provided an ideal, steady dispersion of the nanostructured Mn oxide clusters. In addition, the presence of silica may prevent surface restructuring from deactivating the catalyst's active Mn centers. A year later, Jiao and Boppan [135] reported very efficient and dependable water oxidation catalysts, i.e., MnO_2 nanotubes and MnO_2 nanowires. At low pH, all compounds showed high turnover frequency and strong stability. Navrotsky et al. [133] reported that the oxidation-reduction equilibria of nanophase transition metal oxides showed significant thermodynamically driven shifts. In such variations in oxidation-reduction equilibria at the nanoscale two types of size effects may be important [136]. The first one is dependent on an increased ratio of surface to volume. The unsaturated sites enhanced the system's energy and activated it for many reactions. The second type of effect, true-size effects, also involves changes in local properties. These effects are connected to differences in the electronic characteristics of nano compounds, such as HOMO (highest occupied molecular orbital) or LUMO (lowest unoccupied molecular orbital) compared to bulk compounds (LUMO).

The band gap of nanomaterials can be tuned to absorb in the whole solar spectrum range by varying sizes. The electronic band structure can be controlled by doping. Bottom-up growth approaches allow scalable synthesis of single crystal nanostructures on flexible substrates under mild conditions, leading to light-weight and low cost [137]. Water-splitting efficiency enhancement was observed at the semiconductor/electrolyte interface, where the metal NPs were localized. This is due to the minority charge transport to the electrolyte from the electrode [138]. In an investigation, CdS NPs were prepared with uniform size ranges and experimentally proved improved photocatalytic activities as compared to bulk CdS [139]. The performance of PEC can be enhanced by co-catalyst size effects. One method to increase PEC efficiency is by developing a good catalyst which means a smaller size co-catalyst. Electrokinetics, which causes more electron-hole recombination, dominates in smaller particles. Greater charge extraction at the electrode and electrolyte interface is possible due to the band-bending features of larger particles. Therefore, larger co-catalysts are suitable for improving PEC performance [140].

3.6 SURFACE AREA OF DIFFERENT NANOMATERIALS

The number of active sites for surface redox reactions is significantly increased by the large surface area of nanomaterials. At the electrode/electrolyte interfaces, water splitting requires good surface characteristics and a large surface area [141,142]. Three-dimensional hierarchical TiO_2 BNRs have a high surface area to improve the electrolyte interface, which is good for light harvesting, and a highly efficient electron transport pathway. Due to the strong SPR effect, Au NPs can significantly improve the photocatalytic activity of Au/TiO_2 combination in visible light. Since the surface of Au has a lower Fermi level than TiO_2, a photo-generated electron in TiO$_2$ conduction band may move to the surface of Au, and the recombination of electron-hole pairs occurred [143–147]. Although the electrochemically accessible surface can enhance the reaction volume yet the electrocatalytic reaction is extremely complex and involves a number of interfacial processes. Depending on the crystal facets, electrolysis of water may undergo various adsorption processes even in the case of Pt single crystals [148,149]. Therefore, the charge transfer process at the electrode/electrolyte interface may be controlled by the surface structure at the nano- and atomic scale. For instance, while preparing nanomaterials from graphene, the electrochemical path may be completely different depending on the density of edge or basal planes in the electrochemically accessible areas [150–153]. In a similar way, the PEC process is also dependent on the crystal facet. The presence of a large specific surface area (~2,600 m^2/g), and extended conjugation graphene exhibited higher photocatalytic activity than those of CNTs with a specific surface area (~1,300 m^2/g). The H$_2$ production rates of 29, 175, 19.5, 0.451, and 6 mmol/h/g were observed for the CNT graphene-TiO$_2$ [154], polymer-supported graphene-CdS [155], 2.0.0 Cu$_2$O nanoplatelets oriented on graphene [156], g-C$_3$N$_4$ and graphene composite [157], and nitrogen-doped graphene [118], respectively.

Sun et al. [158] prepared NiCoFe LDH [layered double hydroxide (LDH)] nanoplates on a Co-based nanowire array with a small size, large surface area, and high porosity. As expected, the core/shell nano-array electrode showed enhanced catalytic activity than the corresponding pure LDH NPs and nanowire arrays. Only a small over-potential of 257 mV was required for 80 mA/cm^2 current density. Feng et al. [159] designed a ternary hierarchical composite consisting of Co$_{0.85}$Se nano-sheets, exfoliated graphene foil and NiFe LDH, which had a high surface area of 156 m^2/g and a strong coupling effect. The obtained composite electrode showed a greatly enhanced performance compared to the commercial Ir/C electrode with a high current density of 100 mA/cm^2 at 1.5V vs. RHE. Shi et al. [160] reported NiFe LDH/ graphene with 3.3 times of electrochemical active surface area than the conventional planar materials. This electrode also exposed much higher activity than precious IrO$_2$ electrodes for OER in an alkaline environment. Liu et al. [161] used carbon QDs as support to deposited NiFe LDH nanoplates (3D complex sheet particles). This composite also exhibited excellent electro-activity for OER with an over-potential of ~235 mV to afford the current density of 10 mA/cm^2.

3.7   SURFACE CHARGES OF DIFFERENT NANOMATERIALS

The surface charge of a nanomaterial is an important variable that impacts hydrogen production. Recently, it has been shown that variations in the size and form of the NPs during the production of metal oxide NPs are related to the surface charge density [162,163]. It means the greater the surface charge, the smaller the particle size.

Due to the improved surface separation brought on by Ag NPs, the variations in current density between WO$_3$ and Ag/WO$_3$ at low applied voltage are rather large. However, at high applied voltage ranges, the differences rapidly decrease because the high applied potential mainly favors surface charge separation [164]. This feature may be useful in water splitting for hydrogen production. The serial hole transfer layers of Fe$_2$O$_3$ and NiOOH/FeOOH simultaneously alleviate bulk and surface charge recombination and enhance the water-splitting performance. The results provided theoretical estimation, which showed that the optimized NiOOH /FeOOH/Fe$_2$O$_3$/ BiVO$_4$ photo-anode showed an improved photocurrent of 2.24 mA/cm^2 at 1.23 V vs. RHE, which was near 2.95 times more than that of pristine BiVO$_4$ photo-anode. When compared to BiVO$_4$ photo-anode at 1.23 V vs. RHE, the efficiency of charge transfer in the bulk and surface was increased by 1.63 and 2.62 times, respectively [165].

Since larger surface active sites and shorter charge transport paths both benefit minority carrier extraction and charge transfer kinetics. Nanoscale has been frequently utilized to reduce the size of photocatalysts. However, reducing particle size usually comes at the expense of crystallinity, which increases the density of surface defects and enhances surface recombination. The particle size becomes a complex issue in regulating the photo activities in particulate PEC and PS systems as a result of the competing changes in characteristics brought on by nano-scaling.

REFERENCES

1. Y. Li, J. Zhang, Hydrogen generation from photoelectrochemical water splitting based on nanomaterials, *Laser Photonics Rev.*, 4, 517–528 (2010).
2. J.S. Lee, J. Jang, Hetero-structured semiconductor nanomaterials for photocatalytic applications, *J. Ind. Eng. Chem.*, 20, 363–371 (2014).
3. M. Ni, M.K.H. Leung, D.Y.C. Leung, K. Sumathy, A review and recent developments in photocatalytic water-splitting using TiO2 for hydrogen production, *Renew. Sustain. Energy Rev.*, 11, 401–425 (2007).
4. F. Cao, G. Oskam, G.J. Meyer, P.C. Searson, Electron transport in porous nanocrystalline TiO$_2$ photoelectrochemical cells, *J. Phys. Chem.*, 100, 17021–17027 (1996).
5. E. Stathatos, P. Lianos, Organic/inorganic nanocomposite gels employed as electrolyte supports in Dye-sensitized photoelectrochemical cells, *Int. J. Photoenergy*, 4, 11 (2002).
6. T. Stergiopoulos, I.M. Arabatzis, G. Katsaros, P. Falaras, Binary Polyethylene oxide/ titania solid-state redox electrolyte for highly efficient nanocrystalline TiO2 photoelectrochemical cells, *Nano Lett.*, 2, 1259 (2002).
7. P.R. Mishra, P.K. Shukla, O.N. Srivastava, Study of modular PEC solar cells for photoelectrochemical splitting of water employing nanostructured TiO2 photoelectrodes, *Int. J. Hydrogen Energy* 32, 1680–1685 (2007).
8. H. Zhao, D. Jiang, S. Zhang, W. Wen, Photoelectrocatalytic oxidation of organic compounds at nanoporous TiO2 electrodes in a thin-layer photoelectrochemical cell, *J. Catal.*, 250, 102–109 (2007).
9. D. Chen, Y.F. Gao, G. Wang, H. Zhang, W. Lu, and J.H. Li, Surface tailoring for controlled photoelectrochemical properties: Effect of patterned TiO$_2$ microarrays, *J. Phys. Chem. C*, 111, 13163–13169 (2007).
10. C.J. Lin, Y.T. Lu, C.H. Hsieh, S.H. Chien, Surface modification of highly ordered TiO2 nanotube arrays for efficient photoelectrocatalytic water splitting, *Appl. Phys. Lett.* 94, 113102 (2009).
11. Y.J. Hwang, A. Boukai, P.D. Yang, High density n-Si/n-TiO$_2$ core/shell nanowire arrays with enhanced photoactivity, *Nano Lett.* 9, 410–415 (2009).
12. D. Eder, M. Motta, A.H. Windle, Iron-doped Pt–TiO2 nanotubes for photo-catalytic water splitting, *Nanotechnology*, 20, 055602 (2009).
13. A. Wolcott, W.A. Smith, T.R. Kuykendall, Y.P. Zhao, J.Z. Zhang, Photoelectrochemical water splitting using dense and aligned TiO2 nanorod arrays, *Small*, 5, 104–111 (2009).

14. G.K. Mor, O.K. Varghese, R.H.T. Wilke, S. Sharma, K. Shankar, T.J. Latempa, K.S. Choi, C.A. Grimes, p-Type Cu–Ti–O nanotube arrays and their use in self-biased heterojunction photoelectrochemical diodes for hydrogen generation, *Nano Lett.*, 8, 1906–1911 (2008).

15. X. Cui, M. Ma, W. Zhang, Y.C. Yang, Z.J. Zhang, Nitrogen-doped TiO2 from TiN and its visible light photoelectrochemical properties, *Electrochem. Commun.* 10, 367–371 (2008).

16. J.H. Park, S. Kim, A.J. Bard, Novel carbon-doped TiO2 nanotube arrays with high aspect ratios for efficient solar water splitting, *Nano Lett.* 6, 24–28 (2006).

17. S. Takabayashi, R. Nakamura, Y. Nakato, A nano-modified Si/TiO2 composite electrode for efficient solar water splitting, *J. Photochem. Photobiol. A, Chem.*, 166, 107–113 (2004).

18. S.U.M. Khan, T. Sultana, Photoresponse of n-TiO2 thin film and nanowire electrodes, *Sol. Energy Mater. Sol. Cells* 76, 211–221 (2003).

19. K.R. Reyes-Gil, E.A. Reyes-Garcia, D. Raftery, Nitrogen-doped In2O3 thin film electrodes for photocatalytic water splitting, *J. Phys. Chem. C*, 111, 14579–14588 (2007).

20. R.S. Mane, C.D. Lokhande, photoelectrochemical cells based on nanocrystalline Sb2S3 thin films, *Mater. Chem. Phys.*, 78, 385–392 (2002).

21. R.S. Mane, B.R. Sankapal, C.D. Lokhande, Photoelectrochemical cells based on chemically deposited nanocrystalline Bi2S3 thin films, *Mater. Chem. Phys.*, 60, 196–203 (1999).

22. T. Lindgren, H.L. Wang, N. Beermann, L. Vayssieres, A. Hagfeldt, S.E. Lindquist, Aqueous photoelectrochemistry of hematite nanorod array, *Sol. Energy Mater. Sol. Cells*, 71, 231–243 (2002).

23. S. Saretni-Yarahmadi, K.G.U.Wijayantha, A.A. Tahir, and B. Vaidhyanathan, Nanostructured α-Fe2O3 electrodes for solar driven water splitting: Effect of doping agents on preparation and performance, *J. Phys. Chem. C*, 113, 4768–4778 (2009).

24. K.S. Ahn, S. Shet, T. Deutsch, C.S. Jiang, Y.F. Yan, M. Al- Jassim, J. Turner, Enhancement of photoelectrochemical response by aligned nanorods in ZnO thin films, *J. Power Sources*, 176, 387–392 (2008).

25. K.S. Ahn, Y. Yan, S.H. Lee, T. Deutsch, J. Turner, C.E. Tracy, C.L. Perkins, M. Al-Jassim, Photoelectrochemical properties of N-incorporated ZnO films deposited by reactive RF magnetron sputtering, *J. Electrochem. Soc.*, 154, B956 (2007).

26. I. Bedja, S. Hotchandani, P.V. Kamat, Photoelectrochemistry of quantized WO3 colloids: Electron storage, electrochromic, and photoelectrochromic effects, *J. Phys. Chem.*, 97, 11064–11070 (1993).

27. I. Bedja, S. Hotchandani, R. Carpentier, K. Vinodgopal, P.V. Kamat, Electrochromic and photoelectrochemical behavior of thin WO3 films prepared from quantized colloidal particles, *Thin Solid Films*, 247, 195–200 (1994).

28. I. Saeki, N. Okushi, H. Konno, R. Furuichi, The photoelectrochemical response of TiO2-WO3 mixed oxide films prepared by thermal oxidation of titanium coated with tungsten, *J. Electrochem. Soc.*, 143, 2226 (1996).

29. H.L. Wang, T. Lindgren, J.J. He, A. Hagfeldt, S.E. Lindquist, Photolelectrochemistry of nanostructured WO3 thin film electrodes for water oxidation: Mechanism of electron transport, *J. Phys. Chem. B*, 104, 5686–5696 (2000).

30. U.O. Krasovec, M. Topic, A. Georg, A. Georg, G. Drazic, Preparation and characterisation of nano-structured WO3-TiO2 layers for photoelectrochromic devices, *J. Sol-Gel Sci. Technol.*, 36, 45–52 (2005).

31. A. Wolcott, T.R. Kuykendall, W. Chen, S.W. Chen, J.Z. Zhang, Synthesis and characterization of ultrathin WO3 nanodisks utilizing long-chain poly(ethylene glycol), *J. Phys. Chem. B*, 110, 25288–25296 (2006).

32. X.H. Wang, Chemical characterization of mesoporous material supported ZnO nanoparticles for hydrogen sulfide capture from gas streams, *Adv. Mater. Res.*, 129–131, 143–148, (2010).

33. S.G. Chen, M. Paulose, C. Ruan, G.K. Mor, O.K. Varghese, D. Kouzoudis, C.A. Grimes, Electrochemically synthesized CdS nanoparticle-modified TiO2 nanotube-array photoelectrodes: Preparation, characterization, and application to photoelectrochemical cells, *J. Photochem. Photobiol. A: Chem.*, 177, 177–184 (2006).
34. P. Szymanski, M.A. El-Sayed, Some recent developments in photoelectrochemical water splitting using nanostructured TiO2: A short review, *Theor. Chem. Acc.*, 131, 1202 (2012).
35. T.T. Isimjan, S. Rohani, A.K. Ray, Photoelectrochemical water splitting for hydrogen generation on highly ordered TiO2 nanotubes fabricated by using Ti as cathode, *Int. J Hydrogen Energy*, 37, 103–108 (2012).
36. C. Wang, Q. Hu, J. Huang, L. Wu, Z. Deng, Z. Liu, Efficient hydrogen production by photocatalytic water splitting using N-doped TiO2 film, *Appl. Surf. Sci.*, 283, 188–192 (2013).
37. Z. Zhang, M.F. Hossain, T. Takahashi, Photoelectrochemical water splitting on highly smooth and ordered TiO2 nanotube arrays for hydrogen generation, *Int. J. Hydrogen Energy*, 35, 8528–8535 (2010).
38. M. Kitano, M. Takeuchi, M. Matsuoka, J.M. Thomas, M. Anpo, Photocatalytic water splitting using Pt-loaded visible light responsive TiO2 thin film photocatalysts, *Catal. Today*, 120, 133–138 (2007).
39. A.A. Soliman, H.J. Seguin, Reactively sputtered TiO2 electrodes from metallic targets for water electrolysis using solar energy, *Sol. Energy Mater.*, 5, 95–102 (1981).
40. Y. Liu, L. Tian, X. Tan, X. Li, X. Chen, Synthesis, properties, and applications of black titanium dioxide nanomaterials, *Sci. Bull.*, 62, 431–441 (2017).
41. A. Wolcott, W.A. Smith, T.R. Kuykendall, Y. Zhao, J.Z. Zhang, Photoelectrochemical study of nanostructured ZnO thin films for hydrogen generation from water splitting, *Adv. Funct. Mater.*, 19, 1849–1856 (2009).
42. X. Zhang, Y. Liu, Z. Kang, 3D branched ZnO nanowire arrays decorated with plasmonic Au nanoparticles for high performance photoelectrochemical water splitting, *ACS Appl. Mater. Interfaces*, 6, 4480–4489 (2014).
43. T. Wang, R. Lv, P. Zhang, C. Li, J. Gong, Au nanoparticle sensitized ZnO nanopencil arrays for photoelectrochemical water splitting, *Nanoscale*, 7, 77–81 (2015).
44. C. Zhang, M. Shao, F. Ning, S. Xu, Z. Li, M. Wei, Au nanoparticles sensitized ZnO nanorod@nanoplatelet core-shell arrays for enhanced photoelectrochemical water splitting, *Nano Energy*, 12, 231–239 (2015).
45. Y. Tak, S.J. Hong, J.S. Lee, K. Yong, Solution-based synthesis of a CdS nanoparticle/ZnO nanowire heterostructure array, *Cryst. Growth Des.*, 9, 2627–2632 (2009).
46. G. Wu, M. Tian, A. Chen, Synthesis of CdS quantum-dot sensitized TiO2 nanowires with high photocatalytic activity for water splitting, *J. Photochem. Photobiol. A: Chem.*, 233, 65–71 (2012).
47. J. Zhang, W. Zhu, X. Liu, Stable hydrogen generation from vermiculite sensitized by CdS quantum dot photocatalytic splitting of water under visible-light irradiation, *Dalton Trans.*, 43, 9296–9302 (2014).
48. Y. Dong, R. Wu, P. Jiang, G. Wang, Y. Chen, X. Wu, Efficient photoelectrochemical hydrogen generation from water using a robust photocathode formed by CdTe QDs and nickel ion, *ACS Sustain. Chem. Eng.*, 3, 2429–2434 (2015).
49. Q. Li, H. Meng, P. Zhou, Y. Zheng, J. Wang, J. Yu, Zn1-x CdxS solid solutions with controlled bandgap and enhanced visible-light photocatalytic H2- production activity, *ACS Catal.*, 3, 882–889 (2013).
50. X. Zhou, J. Huang, H. Zhang, H. Sun, W. Tu, Controlled synthesis of CdS nanoparticles and their surface loading with MoS2 for hydrogen evolution under visible light, *Int. J. Hydrogen Energy*, 41, 14758–14767 (2016).
51. Y. Li, L. Wang, T. Cai, S. Zhang, Y. Liu, Y. Song, Glucose-assisted synthesize 1D/2D nearly vertical CdS/MoS2 heterostructures for efficient photocatalytic hydrogen evolution, *Chem. Eng. J.*, 321, 366–374 (2017).

52. Y. Lei, C. Yang, J. Hou, F. Wang, S. Min, X. Ma, Strongly coupled CdS/graphene quantum dots nanohybrids for highly efficient photocatalytic hydrogen evolution: Unraveling the essential roles of graphene quantum dots, *Appl. Catal. B, Environ.*, 216, 59–69 (2017).

53. C.J. Lin, L.C. Kao, Y. Huang, M.A. Banares, S.Y.H. Liou, Uniform deposition of coupled CdS and CdSe quantum dots on ZnO nanorod arrays as electrodes for photoelectrochemical solar water splitting, *Int. J. Hydrogen Energy*, 40, 1388–1393 (2015).

54. S.K. Apte, S.N. Garaje, G.P. Mane, A. Vinu, S.D. Naik, D.P. Amalnerkar, A facile template-free approach for the large-scale solid-phase synthesis of CdS nanostructures and their excellent photocatalytic performance, *Small*, 7, 957–964 (2011).

55. X. Hao, Z. Jin, J. Xu, S. Min, G. Lu, Functionalization of TiO2 with graphene quantum dots for efficient photocatalytic hydrogen evolution, *Superlattices Microstruct*, 94, 237–244 (2016).

56. L. Sang, J. Lin, Y. Zhao, Preparation of carbon dots/TiO2 electrodes and their photoelectrochemical activities for water splitting, *Int. J. Hydrogen Energy*, 42,12122–12132 (2017).

57. C.W. Huang, C.H. Liao, J.C.S. Wu, Y.C. Liu, C.L. Chang, C.H. Wu, Hydrogen generation from photocatalytic water splitting over TiO2 thin film prepared by electron beam-induced deposition, *Int.J. Hydrogen Energy*, 35, 12005–12010 (2010).

58. J. Wang, Z. Wang, Z. Zhu, Synergetic effect of Ni(OH)2 cocatalyst and CNT for high hydrogen generation on CdS quantum dot sensitized TiO2 photocatalyst, *Appl. Catal. B Environ.*, 204, 577–583 (2017).

59. N. Wang, J. Li, L. Wu, X. Li, J. Shu, MnO2 and carbon nanotube co-modified C3N4 composite catalyst for enhanced water splitting activity under visible light irradiation, *Int.J. Hydrogen Energy*, 41, 22743–22750 (2016).

60. X. Li, S. Guo, C. Kan, J. Zhu, T. Tong, S. Ke, Au Multimer@MoS2hybrid structures for efficient photocatalytical hydrogen production via strongly plasmonic coupling effect, *Nano Energy*, 30, 549–558 (2016).

61. T. Puangpetch, T. Sreethawong, S. Yoshikawa, S. Chavadej, Hydrogen production from photocatalytic water splitting over mesoporous-assembled SrTiO3 nanocrystal-based photocatalysts, *J. Mol. Catal. A Chem.*, 312, 97–106 (2009).

62. Y. Zhang, T. Xia, P. Wallenmeyer, C.X. Harris, A.A. Peterson, G.A. Corsiglia, Photocatalytic hydrogen generation from pure water using silicon carbide nanoparticles, *Energy Technol.*, 2, 183–187 (2014).

63. M. Moniruddin, K. Afroz, Y. Shabdan, B. Bizri, N. Nuraje, Hierarchically 3D assembled strontium titanate nanomaterials for water splitting application, *Appl. Surf. Sci.*, 419, 886–892 (2017).

64. D. Akyüz, A. Senocak, B. Köksoy, İ. Ömeroğlu, M. Durmus, E. Demirbas, Coumarin bearing asymmetrical zinc(II) phthalocyanine functionalized SWCNT hybrid nanomaterial: Synthesis, characterization and investigation of bifunctional electrocatalyst behavior for water splitting, *J. Electroanaly. Chem.*, 897, 115552 (2021).

65. M. Fazil, T. Ahmad, Pristine TiO2 and Sr-Doped TiO2 Nanostructures for enhanced photocatalytic and electrocatalytic water splitting applications, *Catalysts*, 13, 93 (2023).

66. J. Li, K. Zhao, Y. Zhang, Facet-level mechanistic Insights into general homogeneous carbon doping for enhanced solar-to-hydrogen conversion, *Adv. Funct. Mater.*, 25, 2189–2201 (2015).

67. L. Ye, X. Jin, Y. Leng, Y. Su, H. Xie, C. Liu, Synthesis of black ultrathin BiOCl nanosheets for efficient photocatalytic H2 production under visible light irradiation, *J. Power Sources*, 293, 409–415 (2015).

68. L. Zhang, Z.K. Han, W.Z. Wang, X. Li, Y. Su, D. Jiang, X.L. Lei, Solar-light-driven pure water splitting with ultrathin BiOCl nanosheets, *Chemistry*, 21,18089–18094 (2015).

69. H. Khan, I.H. Lone, S.E. Lofland, K.V. Ramanujachary, T. Ahmad, Exploiting multiferroicity of TbFeO3 nanoparticles for hydrogen generation through photo/electro/photoelectro-catalytic water splitting, *Int. J. Hydrogen Energy*, 48, 5493–5505 (2023).

70. H. Khan, J. Ahmed, S.E. Lofland, K.V. Ramanujachary, T. Ahmad, Synergistic effect of multiferroicity in GdFeO3 nanoparticles for significant hydrogen production through photo/electrocatalysis, *Mater. Today Chem.*, 33, 101713 (2023).

71. H. Khan, S.E. Lofland, K.V. Ramanujachary, N.Alhokbany, T. Ahmad, Bifunctional multiferroic GdCrO$_3$ nanoassemblies for sustainable H$_2$ production using electro- and photocatalysis, *ACS Appl. Energy Mater.*, 6 (15), 8102–8110 (2023).

72. M. Fazil, S.M. Alshehri, Y. Mao, T. Ahmad, Hydrothermally derived Mg-doped TiO2 nanostructures for enhanced H2 evolution using photo- and electro-catalytic water splitting, *Catalysts*, 13, 89 (2023).

73. F. Naaz, T. Ahmad, Ag-doped WO3 nanoplates as heterogenous multifunctional catalyst for glycerol acetylation, electrocatalytic and enhanced photocatalytic hydrogen production, *Langmuir*, 39 (27), 9300–9314 (2023).

74. S.A. Ali, J. Ahmed, Y. Mao, T. Ahmad, Symbiotic MoO3–SrTiO3 heterostructured nanocatalysts for sustainable hydrogen energy: Combined experimental and theoretical simulations, *Langmuir*, 39(36), 12692–12706 (2023).

75. A.T. Harris, R. Bali, On the formation and extent of uptake of silver nanoparticles by live plants, *J. Nanopart. Res.*, 10, 691–695 (2008).

76. S. Ghosh, S. Patil, M. Ahire, R. Kitture, D.D. Gurav, A.M. Jabgunde, S. Kale, K. Pardesi, V. Shinde, J. Bellare, D.D. Dhavale, B.A. Chopade Gnidia glauca flower extract mediated synthesis of gold nanoparticles and evaluation of its chemocatalytic potential, *J. Nanobiotechnol.*, 10, 17 (2012).

77. M. Khan, S.F. Adil, M.N. Tahir, W. Tremel, H.Z. Alkhathlan, A. Al-Warthan, M.R. Siddiqui, Green synthesis of silver nanoparticles mediated by Pulicaria glutinosa extract, *Int. J. Nanomed.*, 8, 1507–1516 (2013).

78. M. Rai, A. Yadav, Plants as potential synthesiser of precious metal nanoparticles: Progress and prospects, *IET Nanobiotechnol.*, 7, 117–124 (2013).

79. Y. Qu, X. Pei, W. Shen, X. Zhang, J. Wang, Z. Zhang, S. Li, S. You, F. Ma, J. Zhou, Biosynthesis of gold nanoparticles by Aspergillum sp. WL-Au for degradation of aromatic pollutants, *Phys. E. Low-dimens. Syst. Nanostruct.*, 88, 133–141 (2017).

80. M. Martins, C. Mourato, S. Sanches, J.P. Noronha, M.B. Crespo, I.A. Pereira, Biogenic platinum and palladium nanoparticles as new catalysts for the removal of pharmaceutical compounds, *Water Res.*, 108, 160–168 (2017).

81. L. Yue, J. Wang, Y. Zhang, S. Qi, B. Xin, Controllable biosynthesis of high-purity lead-sulfide (PbS) nanocrystals by regulating the concentration of polyethylene glycol in microbial system, *Bioproc. Biosyst. Eng.*, 39, 1839–1846 (2016).

82. C.S. De, T. Hennebel, G.B. De, W. Verstraete, N. Boon, Biopalladium: From metal recovery to catalytic applications, *Microb. Biotechnol.*, 5, 5–17 (2012).

83. T. Hennebel, C.S. De, L. Vanhaecke, K. Vanherck, I. Forrez, B. Gusseme, P. Verhagen, K. Verbeken, D.B.B. Van, I. Vankelecom, N. Boon, Removal of diatrizoate with catalytically active membranes incorporating microbially produced palladium nanoparticles, *Water Res.*, 44, 1498–1506 (2010).

84. C.P. Raja, J.M. Jacob, R.M. Balakrishnan, Selenium biosorption and recovery by marine Aspergillus terreus in an upflow bioreactor, *J. Environ. Eng.*, 142, (2015).

85. J. Tu, Z. Yang, C. Hu, Efficient catalytic aerobic oxidation of chlorinated phenols with mixed-valent manganese oxide nanoparticles, *J. Chem. Technol. Biotechnol.*, 9, 80–86 (2015).

86. X. Xiao, X.B. Ma, H. Yuan, P.C. Liu, Y.B. Lei, H. Xu, D.L. Du, J.F. Sun, Y.J. Feng, Photocatalytic properties of zinc sulfide nanocrystals biofabricated by metal-reducing bacterium Shewanella oneidensis MR-1, *J. Hazard Mater.*, 288, 134–139 (2015).

87. E. Ismail, A. Diallo, M. Khenfouch, S.M. Dhlamini, M. Maaza, RuO2 nanoparticles by a novel Green process via Aspalathus linearis natural extract & their water splitting response, *J. Alloys Comp.*, 662, 283–289 (2016).
88. A. Realpe, D.P. Nunez, A. Herrera, Synthesis of Fe-TiO2 nanoparticles for photoelectrochemical generation of hydrogen, *Int. J. Chem. Tech.*, 9, 453–464 (2016).
89. D.V. Raorane, P.S. Chavan, S.R. Pednekar, R.S. Chaughule, Green and rapid synthesis of copper-doped TiO2 nanoparticles with increased photocatalytic activity, *Adv. Chem. Sci.*, 6, 13–20 (2017).
90. M.F. Alajmi, J. Ahmed, A. Hussain, T. Ahamad, N. Alhokbany, S. Amir, T. Ahmad, S.M. Alshehri, Green synthesis of Fe3O4 nanoparticles using aqueous extracts of Pandanus odoratissimus leaves for efficient bifunctional electrocatalytic activity, *Appl. Nanosci.*, 8, 1427–1435 (2018).
91. V. Selvanathan, M. Shahinuzzaman, S. Selvanathan, D.K. Sarkar, N. Algethami, H.I. Alkhammash, F.H. Anuar, Z. Zainuddin, M. Aminuzzaman, H. Abdullah, M. Akhtaruzzaman, Phytochemical-assisted Green synthesis of nickel oxide nanoparticles for application as electrocatalysts in oxygen evolution reaction, *Catalysts*, 11, 1523 (2021).
92. T. Zahra, K.S. Ahmad, C. Zequine, R.K. Gupta, A.G. Thomas, M.A. Malik, Evaluation of electrochemical properties for water splitting by NiO nano-cubes synthesized using Olea ferruginea Royle, *Sustain. Energy Technol. Assess.*, 40, 100753 (2020).
93. X.F. Zhang, Q.J. Wu, Z. Du, Y. Zheng, Q.B. Li, Green synthesis of g-C3N4-Pt catalyst and application to photocatalytic hydrogen evolution from water splitting, *Fullerenes, Nanotubes Carbon Nanostruct.*, 26, 688–695 (2018).
94. T. Zahra, K.S. Ahmad, C. Zequine, R.K. Gupta, A.G. Thomas, M.A. Malik, S.B. Jaffri, D. Ali, Electro-catalyst [ZrO2/ZnO/PdO]-NPs green functionalization: Fabrication, characterization and water splitting potential assessment, *Int. J. Hydrogen Energy*, 46, 19347–19362 (2021).
95. M. Ayub, M.H.D. Othman, M. Zamri, M. Yusop, I.U. Khan, Z.S. Tai, S.K. Hubadillah, Optimized single-step synthesis of graphene-based carbon nanosheets from palm oil fuel ash, *Mater. Chem. Phys.*, 296, 127202, (2023).
96. T.H. Bointon, M.D. Barnes, S. Russo, M.F. Craciun, High quality monolayer graphene synthesized by resistive heating cold wall chemical vapor deposition, *Adv. Mater.*, 27, 4200–4206 (2015).
97. M.T. Khorshid, E. Omrani, P.L. Menezes, P.K. Rohatgi, Tribological performance of self-lubricating aluminum matrix nanocomposites: Role of graphene nanoplatelets, *Eng. Sci. Technol. Int. J.*, 19, 463–469 (2016).
98. X. Li, W.C.H. Choy, X. Ren, D. Zhang, H. Lu, Highly intensified surface enhanced Raman scattering by using monolayer graphene as the nanospacer of metal film-metal nanoparticle coupling system, *Adv. Funct. Mater.*, 24, 3114–3122 (2014).
99. A. Politano, G. Chiarello, Alkali-induced hydrogenation of epitaxial graphene by water splitting at 100 K, *J. Chem. Phys.*, 138, 004470 (2013).
100. T.F. Yeh, J. Cihlar, C.Y. Chang, C. Cheng, H. Teng, Roles of graphene oxide in photocatalytic water splitting, *Mater. Today*, 16, 78–84 (2013).
101. J. Yang, X. Zeng, L. Chen, W. Yuan, Photocatalytic water splitting to hydrogen production of reduced graphene oxide/SiC under visible light, *Appl. Phys. Lett.*, 102, 083101 (2013).
102. K. Iwashina, A. Iwase, Y.H. Ng, R. Amal, A. Kudo, Z-schematic water splitting into H2 and O2 using metal sulfide as a hydrogen-evolving photocatalyst and reduced graphene oxide as a solid-state electron mediator, *J. Am. Chem. Soc.*, 137, 604–607 (2015).
103. S. Singla, S. Sharma, S. Basu, N.P. Shetti, K.R. Reddy, Graphene/graphitic carbon nitride-based ternary nanohybrids: Synthesis methods, properties, and applications for photocatalytic hydrogen production, *FlatChem*, 24, 100200 (2020).

104. H. Liu, Y. Liu, D. Zhu. Chemical doping of graphene, *J. Mater. Chem.*, 21, 3335–3345 (2011).
105. S. Wei, R. Zhang, Y. Liu, H. Ding, Y.L. Zhang. Graphene quantum dots prepared from chemical exfoliation of multiwall carbon nanotubes: An efficient photocatalyst promoter, *Catal. Commun.*, 74, 104–109 (2016).
106. D.J. Martin, G. Liu, S.J.A. Moniz, Y. Bi, A.M. Beale, J. Ye, J.W. Tang, Efficient visible driven photocatalyst, silver phosphate: Performance, understanding and perspective, *Chem. Soc. Rev.*, 44, 7808–7828 (2015).
107. Z. Sheng, L. Shao, J. Chen, W. Bao, F. Wang, X. Xia, Catalyst-free synthesis of nitrogen-doped graphene via thermal annealing graphite oxide with melamine and its excellent electrocatalysis, *ACS Nano*, 5, 4350–4358 (2011).
108. S. Kawai, S. Saito, S. Osumi, S. Yamaguchi, A.S. Foster, P. Spijker, F. Meyer, Atomically controlled substitutional boron doping of graphene nanoribbons, *Nat. Commun.*, 6, 1–6 (2015).
109. A. Kimouche, M.M. Ervasti, R. Drost, S. Halonen, A. Harju, P.M. Joensuu, J. Sainio, P. Liljeroth, Ultra-narrow metallic armchair graphene nanoribbons, *Nat, Commun.*, 6, 10177 (2015).
110. T.D. Nguyen-Phan, S. Luo, Z. Liu, A.D. Gamalski, J. Tao, W. Xu, E.A. Stach, D.E. Polyansky, S.D. Senanayake, E. Fujita, J.A. Rodriguez, Striving toward noble-metal-free photocatalytic water splitting: The hydrogenated-graphene–TiO2 prototype, *Chem. Mater.*, 27, 6282–6296 (2015).
111. S. Pei, Q. Wei, K. Huang, H.M. Cheng, W. Ren, Green synthesis of graphene oxide by seconds timescale water electrolytic oxidation, *Nat. Commun.*, 9, 1–9 (2018).
112. S. Son, J.M. Lee, S.J. Kim, H. Kim, X. Jin, K.K. Wang, M. Kim, J.W. Hwang, W. Choi, Y.R. Kim, Understanding the relative efficacies and versatile roles of 2D conductive nanosheets in hybrid-type photocatalyst, *Appl. Catal. B Environ.*, 257, 117875 (2019).
113. C. Dong, X. Li, P. Jin, W. Zhao, J. Chu, J. Qi, Intersubunit electron transfer (IET) in quantum dots/graphene complex: What features does IET endow the complex with? *J. Phys. Chem. C*, 116, 15833–15838 (2012).
114. H. Xu, L. Ma, Z. Jin, Nitrogen-doped graphene: Synthesis, characterizations and energy applications, *J. Energy Chem.*, 27, 146–160 (2018).
115. H. Zhang, X. Lv, Y. Li, Y. Wang, J. Li, P25-graphene composite as a high performance photocatalyst, *ACS Nano*, 4, 380–386 (2010).
116. X.Y. Zhang, H.P. Li, X.L. Cui, Y. Lin, Graphene/TiO2 nanocomposites: Synthesis, characterization and application in hydrogen evolution from water photocatalytic splitting, *J. Mater. Chem.*, 20, 2801–2806 (2010).
117. C. Lavorato, A. Primo, R. Molinari, H. Garcia, N-doped graphene derived from biomass as a visible-light photocatalyst for hydrogen generation from water/methanol mixtures, *Chem. A Eur. J.*, 20, 187–194(2014).
118. T.F. Yeh, C.Y. Teng, S.J. Chen, H. Teng, Nitrogen-doped graphene oxide quantum dots as photocatalysts for overall watersplitting under visible light Illumination, *Adv. Mater.*, 26, 3297–3303 (2014).
119. D. Zhang, B. Hu, D. Guan, Z. Luo, Essential roles of defects in pure graphene/Cu2O photocatalyst, *Catal. Commun.*, 76, 7–12 (2016).
120. D. Mateo, I. Esteve-Adell, J. Albero, A. Primo, H. Garcia, Oriented 2.0. 0 Cu2O nano-platelets supported on few-layers graphene as efficient visible light photocatalyst for overall water splitting, *Appl. Catal. B Environ.*, 201, 582–590 (2017).
121. Yi. Park, S.H. Kanga, W. Choi, Exfoliated and reorganized graphite oxide on titania nanoparticles as an auxiliary co-catalyst for photocatalytic solar conversion, *Phys. Chem. Chem. Phys.*, 13, 9425–9431 (2011).
122. Q. Xiang, J. Yu, M. Jaronie, Enhanced photocatalytic H2-production activity of graphene-modified titaniananosheets, *Nanoscale*, 3, 3670–3678 (2011).

123. M.J. Rivero, O. Iglesias, P. Ribao, I. Ortiz, Kinetic performance of TiO2/Pt/reduced graphene oxide composites in the photocatalytic hydrogen production, *Int. J. Hydrogen Energy*, 44, 101–109 (2019).

124. A. Ye, W. Fan, Q. Zhang, W. Deng, Y. Wang, CdS-graphene and CdS-CNT nanocomposites as visible-light photocatalysts for hydrogen evolution and organic dye degradation, *Catal. Sci. Technol.*, 2, 969–978 (2012).

125. Z. Sun, J. Guo, S. Zhu, L. Mao, J. Ma, D. Zhang, A high-performance Bi2WO6–graphene photocatalyst for visible light-induced H2 and O2 generation, *Nanoscale*, 6, 2186–2193 (2014).

126. Q. Xiang, J. Yu, M. Jaroniec, Synergetic effect of MoS2 and graphene as cocatalysts for enhanced photocatalytic H2 production activity of TiO2 nanoparticles, *J. Am. Chem. Soc.*, 134, 6575–6578 (2012).

127. F. Wang, M. Zheng, C. Zhu, B. Zhang, W. Chen, L. Ma, W. Shen, Visible light photocatalytic H2-production activity of wide band gap ZnS nanoparticles based on the photosensitization of graphene, *Nanotechnology*, 26, 345402 (2015).

128. G. Nagaraju, K. Manjunath, S. Sarkar, E. Gunter, S.R. Teixeira, J. Dupont, TiO2-RGO hybrid nanomaterials for enhanced water splitting reaction, *Int. J. Hydrogen Energy*, 40, 12209–12216 (2015).

129. T. Lv, M. Wu, M. Guo, Q. Liu, L. Jia, Self-assembly photocatalytic reduction synthesis of graphene encapsulated LaNiO3 nanoreactor with high efficiency and stability for photocatalytic water splitting to hydrogen, *Chem. Eng. J.*, 356, 580–591 (2019).

130. P. Wang, S. Zhan, Y. Xia, S. Ma, Q. Zhou, Y. Li, The fundamental role and mechanism of reduced graphene oxide in rGO/Pt- TiO2 nanocomposite for high-performance photocatalytic water splitting, *Appl. Catal. B Environ.*, 207, 335–346 (2017).

131. J. Zhang, L. Qi, J. Ran, J. Yu, S.Z. Qiao, Ternary NiS/ZnxCd1xS/reduced graphene oxide nanocomposites for enhanced solar photocatalytic H2-production activity, *Adv. Energy Mater.*, 4, 1301925 (2014).

132. J. Maier, Thermodynamic aspects and morphology of nano-structured ion conductors: Aspects of nano-ionics Part I, *Solid State Ionics*, 154, 291–301 (2002).

133. A. Navrotsky, C. Ma, K. Lilova, N. Birkner, Nanophase transition metal oxides show large thermodynamically driven shifts in oxidation-reduction equilibria, *Science*, 330, 199–201 (2010).

134. F. Jiao, H. Frei, Nanostructured manganese oxide clusters supported on mesoporous silica as efficient oxygen-evolving catalysts, *Chem. Commun.*, 46, 2920–2922 (2010).

135. V.B.R. Boppana, F. Jiao, Nanostructured MnO2: An efficient and robust water oxidation catalyst, *Chem. Commun.*, 47, 8973–8975 (2011).

136. M.M. Najafpour, F. Rahimi, E.M. Aro, C.H. Lee, S.I. Allakhverdiev, Nano-sized manganese oxides as biomimetic catalysts for water oxidation in artificial photosynthesis: A review, *J. Roy. Soc. Inter.*, 9, 2383–2395 (2012).

137. J.Z. Zhang, Metal oxide nanomaterials for solar hydrogen generation from photoelectrochemical water splitting, *MRS Bull.*, 36, 48–55 (2011).

138. Y. Zhong, K. Ueno, Y. Mori, X. Shi, T. Oshikiri, K. Murakoshi, Plasmon-assisted water splitting using two sides of the same SrTiO3 single-crystal substrate: Conversion of visible light to chemical energy, *Angew Chemie Int. Ed.*, 53, 10350–10354 (2014).

139. M. Sathish, B. Viswanathan, R.P. Viswanath, Alternate synthetic strategy for the preparation of CdS nanoparticles and its exploitation for water splitting, *Int. J. Hydrogen Energy*, 31, 891–898 (2006).

140. S. Pokrant, S. Dilger, S. Landsmann, M. Trottmann, Size effects of cocatalysts in photoelectrochemical and photocatalytic water splitting, *Mater. Today Energy*, 5, 158–163 (2017).

141. Y. Kuang, T. Yamada, K. Domen, Surface and interface engineering for photoelectrochemical water oxidation, *Joule*, 1, 290–305, (2017).

142. N. Guijarro, M.S. Prevot, K. Sivula, Surface modification of semiconductor photoelectrodes, *Phys. Chem. Chem. Phys.*, 17, 15655–15674 (2015).

143. C.G. Silva, R. Juarez, T. Marino, R. Molinari, H. Garcia, Influence of excitation wavelength (UV or visible light) on the photocatalytic activity of titania containing gold nanoparticles for the generation of hydrogen or oxygen from water, *J. Am. Chem. Soc.*, 133, 595–602 (2011).

144. Z. Zhang, L. Zhang, M.N. Hedhili, H. Zhang P. Wang, Plasmonic gold nanocrystals coupled with photonic crystal seamlessly on TiO2 nanotube photoelectrodes for efficient visible light photoelectrochemical water splitting, *Nano Lett.*, 13, 14–20 (2020).

145. J.J. Chen, J.C.S. Wu, P.C. Wu, D.P. Tsai, Plasmonic photocatalyst for H2 evolution in photocatalytic water splitting, *J. Phys. Chem. C*, 115, 210–216 (2011).

146. A. Tanaka, S. Sakaguchi, K. Hashimoto, H. Kominami, Preparation of Au/TiO2 with metal cocatalysts exhibiting strong surface plasmon resonance effective for photo-induced hydrogen formation under irradiation of visible light, *ACS Catal.*, 3, 79–85 (2012).

147. Z. Liu, W. Hou, P. Pavaskar, M. Aykol, S.B. Cronin, Plasmon resonant enhancement of photocatalytic water splitting under visible illumination, *Nano Lett.*, 11, 1111–1116 (2011).

148. J. Barber, S. Morin, B. Conway, Specificity of the kinetics of H2 evolution to the structure of single-crystal pt surfaces, and the relation between Opd and Upd H, *J. Electroanal. Chem.*, 446, 125–138 (1998).

149. V. Stamenkovic, N.M. Markovic, P. Ross, Structure relationships in electrocatalysis: Oxygen reduction and hydrogen oxidation reactions on Pt(111) and Pt(100) in solutions containing chloride ions, *J. Electroanal. Chem.*, 500, 44–51 (2001).

150. A. Eftekhari, B. Yazdani, Initiating electropolymerization on graphene sheets in graphite oxide structure, *J. Polym. Sci., Part A Polym. Chem.*, 48, 2204–13 (2010).

151. C.E. Banks, T.J. Davies, G.G. Wildgoose, R.G. Compton, Electrocatalysis at graphite and carbon nanotube modified electrodes: Edge-plane sites and tube ends are the reactive sites, *Chem. Commun.*, 829–841 (2005).

152. A. Eftekhari, H. Garcia, The necessity of structural irregularities for the chemical applications of graphene, *Mater. Today Chem.*, 4, 1–16 (2017).

153. W. Yuan, Y. Zhou, Y. Li, C. Li, H. Peng, J. Zhang, Z. Liu, L. Dai, G. Shi, The edge and basal-plane-specific electrochemistry of a single-layer graphene sheet, *Sci. Rep.*, 3, 02248 (2013).

154. S. Bellamkonda, N. Thangavel, H.Y. Hafeez, B. Neppolian, G.R. Rao, Highly active and stable multi-walled carbon nanotubes-graphene-TiO2 nanohybrid: An efficient nonnoble metal photocatalyst for water splitting, *Catal. Today*, 120–127 (2019).

155. J. Xu, L. Wang, X. Cao, Polymer supported graphene-CdS composite catalyst with enhanced photocatalytic hydrogen production from water splitting under visible light, *Chem. Eng. J.*, 283, 816–825 (2016).

156. D. Mateo, I. Esteve-Adell, J. Albero, A. Primo, H. Garcia, Oriented 2.0.0 Cu2O nanoplatelets supported on few-layers graphene as efficient visible light photocatalyst for overall water splitting, *Appl. Catal. B Environ.*, 201, 582–590 (2017).

157. Q. Xiang, J. Yu, M. Jaroniec, Preparation and enhanced visible light photocatalytic H$_2$-production activity of graphene/C$_3$N$_4$ composites, *J. Phys. Chem. C*, 115, 7355–7363 (2011).

158. X. Sun, Q. Yang, T. Li, Z. Lu, J. Liu, Hierarchical construction of an ultrathin layered double hydroxide nanoarray for highly-efficient oxygen evolution reaction. *Nanoscale*, 6, 11789–11794 (2014).

159. Y. Hou, M.R. Lohe, J. Zhang, S. Liu, X. Zhuang, X. Feng, Vertically oriented cobalt selenide/NiFe layered-double-hydroxide nanosheets supported on exfoliated graphene foil: An efficient 3D electrode for overall water splitting, *Energy Environ. Sci.*, 9, 478–483(2016).

160. G. Shi, X. Yu, M. Zhang, W. Yuan, A high-performance three-dimensional Ni–Fe layered double hydroxide/graphene electrode for water oxidation, *J. Mater. Chem. A*, 3, 6921–6928 (2015).
161. Y. Liu, D. Tang, X. Wu, R. Liu, X. Han, Y. Han, H. Huang, J. Liu, Z. Kang, Carbon quantum dot/NiFe layered double-hydroxide composite as a highly efficient electrocatalyst for water oxidation, *ACS Appl. Mater. Interfaces*, 6, 7918–7925 (2014).
162. L. Vayssieres, C. Chaneac, E. Tronc, J.P. Jolivet, Size tailoring of magnetite particles formed by aqueous precipitation: An example of thermodynamic stability of nanometric oxide particles, *J. Colloid Interface Sci.*, 1998, 205, 205–212 (1998).
163. A. Pottier, S. Cassaigon, C. Chaneac, F. Villan, E. Tronc, J.P. Jolivet, Size tailoring of TiO2 anatase nanoparticles in aqueous medium and synthesis of nanocomposites. Characterization by Raman spectroscopy, *J. Mater. Chem.*, 13, 877–882 (2003).
164. Y. Li, W. Zhang, B. Qiu, Enhanced surface charge separation induced by Ag nanoparticles on WO3 photoanode for photoelectrochemical water splitting, *Chem. Lett.*, 49, 741–744 (2020).
165. L. Li, J. Li, J. Bai, Q. Zeng, L. Xia, Y. Zhang, S. Chen, Q. Xu, B. Zhou, Serial hole transfer layers for BiVO4 photoanode with enhanced photoelectrochemical water splitting, *Nanoscale*, 10, 18378–18386 (2018).

4 Photo water splitting for hydrogen production

4.1 INTRODUCTION

Innovative technologies are being investigated as a result of the global search for clean and sustainable energy sources. Of these, photo water splitting appears to be a potential route toward the creation of renewable hydrogen. This method uses the energy of sunshine to catalyze the dissolution of water into its constituent elements, oxygen and hydrogen. A semiconductor (SC) material designed to capture sunlight and start the water-splitting processes lies at the heart of this device. One of photo water splitting's benefits is that it could be a clean, sustainable way to produce hydrogen. The development of effective and commercially feasible photo water-splitting technology could be crucial in accelerating the shift to a more sustainable and environmentally friendly energy landscape as long as research and development activities are sustained.

The fact that photo water-splitting methods rely on freely available solar energy is what makes them so important. The sun emits some energy on a clear day, with about 1,361 watts per square meter (W/m^2) of solar irradiance reaching the Earth's outer atmosphere. The solar constant is the name given to this variable. However, the atmosphere scatters, absorbs, and reflects solar energy before it reaches the Earth's surface, lowering the amount of sunlight that actually reaches the ground. Under clear skies, the average solar irradiation at the Earth's surface is around 1,000 W/m^2. Good semiconductors can harvest this amount of energy.

There are different types of photo water-splitting techniques depending on the working principles. The most important are photo-thermal, photo-electrochemical (PEC), photo-electrolysis, and direct photocatalytic methods. Although water radiolysis does not depend on sun energy, yet we have included water radiolysis in this chapter. The reason is that water radiolysis depends on the application of high energetic radiations. These methods are discussed in detail in this chapter so that academicians, researchers, industrial people, and government authorities can use them.

4.2 PHOTO-THERMAL METHOD

The photo-thermal method for water splitting is a technique used to harness solar energy to separate water into its constituent elements, hydrogen and oxygen. In this process, heat produced by sunlight is used to propel chemical reactions that split water molecules into hydrogen and oxygen. A substance that is capable of effectively absorbing sunlight and converting it into heat is essential to this process. The fundamental principles of the photo-thermal method of water splitting are dependent upon

DOI: 10.1201/9781003432364-5

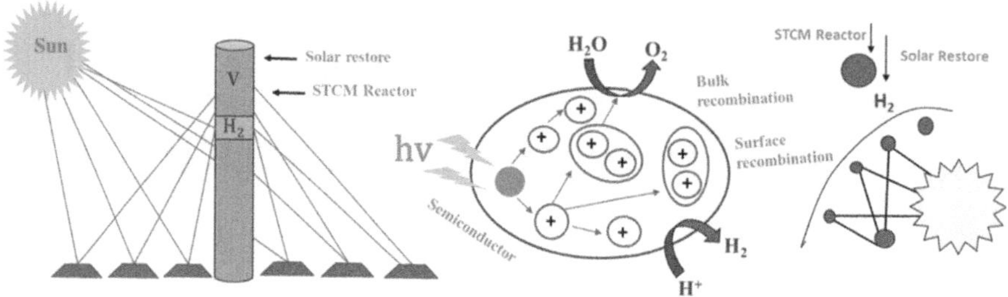

FIGURE 4.1 Photo-generation of charge carriers in a semiconductor (SC) and possible ways of their distribution.

the absorption of light, the production of heat, the vaporization of water, the splitting of water, and the collection of products [1].

A light-absorbing material, frequently an SC, is exposed to sunlight. The SC absorbs the solar energy and changes it into heat. The SC absorbs sunlight and heats up, rising its temperature meaningfully. Water is entered into the hot material, and the high temperature origins the water to vaporize. The water vapor then reaches the hot surface. The high-temperature water vapor experiences a thermochemical reaction, which breaks water molecules (H_2O) into hydrogen (H_2) and oxygen (O_2). This reaction characteristically comprises a metal oxide, like ferrite or cerium oxide, that acts as a catalyst. The produced hydrogen and oxygen gases are collected and used for numerous purposes, including clean energy storage or as a feedstock for chemical processes [2].

Some reactors have been developed and are used. The most important are solar, infrared, and microwave reactors. These depend on the different types of radiations as clear from their names. The necessary high temperatures can be generated by concentrating sunlight onto a reactor tower using a field of mirror "heliostats" (Figure 4.1) [3,4].

The chemicals used in the process are reused within each cycle, creating a closed loop that consumes only water and produces hydrogen and oxygen. In this method of water-splitting process, metal oxides are used as catalysts. The metal oxide is converted first into a reduced form following the splitting of water as below.

$$MO_{oxd} \rightarrow MO_{redn} + \tfrac{1}{2}O_2 \tag{4.1}$$

$$MO_{redn} + H_2O \rightarrow MO_{oxd} + H_2 \tag{4.2}$$

The other important aspects of this method are thermodynamics, temperature swing vs. pressure swing cycling, two-step thermochemical water splitting, and volatile and non-volatile redox materials. [5]. This method is effective but needs a lot of energy, which is difficult to achieve by solar radiation in an inexpensive way. Therefore, there is a great demand to improve the performance and explore new methods.

One of the benefits of the photo-thermal water splitting is operating at high temperatures that can lead to additional effective hydrogen manufacture. Nevertheless, it also needs advanced materials and regulated systems to accomplish the high temperatures and optimize the change of solar energy into chemical energy. Furthermore, the efficacy of this method depends on the choice of materials and reactor strategy. Investigators continue to examine and develop photo-thermal water-splitting methods as part of the ongoing strength to find sustainable and effective ways to yield hydrogen as a clean energy source. These approaches have the talent to play a role in renewable energy systems and the manufacture of green hydrogen.

4.3 PHOTO-ELECTROCHEMICAL (PEC) METHOD

The PEC water-splitting method is a process that utilizes light and specific materials to directly transform solar energy into chemical energy by splitting water into hydrogen and oxygen. It contains principles from both photochemistry and electrochemistry to attain this change. The main component in a PEC cell is a photo-electrode, which absorbs light and brings the water-splitting reaction. This process includes photo-electrode, light absorption, electron transfer, hydrogen and oxygen evolution, their collection, and electric circuit. A schematic representation of PEC is shown in Figure 4.2.

The various types of photocatalysts (semiconductors) are used in this procedure. The most significant are TiO_2 [6], WO_3 [7,8], Fe_2O_3 [9,10], and $BiVO_4$ [11–13]. These materials are good due to their suitable band gaps, good chemical inertness, greater optical steadiness, cheapness, and photo-stability [14–16]. The photo-electrode is located in a water electrolyte solution. When the photo-electrode is open to sunlight, it absorbs photons and creates electron-hole pairs, making electron-rich and electron-deficient areas within the material. The electron-rich area contributes electrons to the water in the electrolyte, starting the reduction of water to yield hydrogen ions (H^+). The electron-deficient area receives electrons, which contribute in the oxidation of water, leading to the development of oxygen and protons. Hydrogen ions travel to a separate electrode called the cathode, where they receive electrons and form hydrogen gas (H_2). Oxygen gas (O_2) is produced at another electrode, naturally

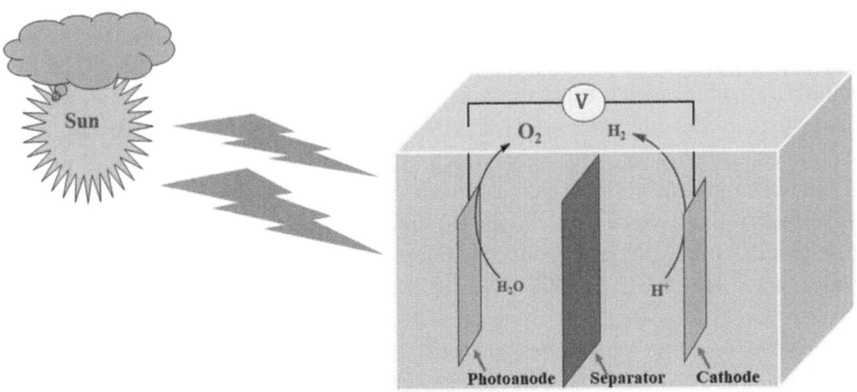

FIGURE 4.2   A schematic representation of the photo-electrochemical (PEC) method.

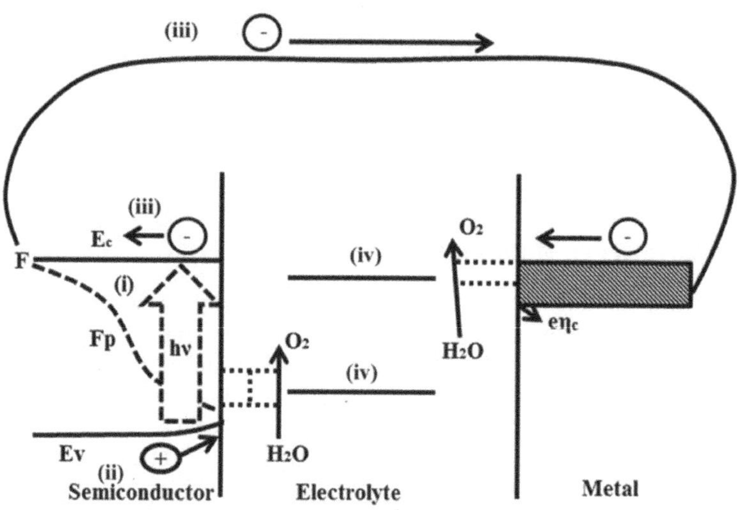

FIGURE 4.3 Basic diagram for a photo-anode (n-type semiconductor); (i) absorption of light, (ii) transfer of charge; (iii) charge carrier transportation; and (iv) chemical reactions at surface [reproduced with permission from ref. 17].

the anode. To sustain the flow of electrons and ions, an outside electrical circuit is established between the photo-electrode, $\rightarrow$ cathode, and anode. At the cathode, the reduction reaction occurs ($2H^+ + 2e\ H_2$) and at $\rightarrow$ the photo-anode the reaction of O_2 evolution is done ($2H_2O\ 4H^+ + O_2 + 4e^-$). The free energy change for the conversion of one water molecule into one molecule of hydrogen and half molecule of oxygen in normal situations is $\Delta G = 237.2$ kJ/mol, which agrees to $\Delta E° = 1.23$ V per transported electron, as per the Nernst equation. Consequently, the semiconductor should absorb photons with energy 1.23 eV and convert water molecules into H_2 and O_2. The actual progressions taking place in the PEC processes are designated in Figure 4.3 [17].

Wenderich et al. [18] reported the industrial feasibility of PEC water-splitting hydrogen production. The authors described that the PEC method is competing with other industrial methods of hydrogen production. Furthermore, the authors emphasized the need of more research for the advancement of this method. The PEC method provides numerous advantages like direct conversion, renewable energy, clean hydrogen production, and versatility. The PEC method produces hydrogen and oxygen directly from sunlight, which makes it a productive method of producing these important gases. It is a clean and sustainable technology because it uses solar energy that is renewable. Since no greenhouse gases are released during the PEC process, the hydrogen produced is referred to as green hydrogen. As photoelectrodes, a variety of SC materials can be employed, enabling customization based on budget and efficiency requirements. The PEC method does come with certain drawbacks, including the requirement for stable and effective photoelectrode materials, efficient electrolytes, and PEC cell designs to enhance the overall performance. Scientists are presently engaged in the development of sophisticated materials and PEC optimization.

4.4 PHOTO-ELECTROLYSIS METHOD

Photo-electrolysis and photochemical water splitting are two separate approaches for using light energy to break water into its hydrogen and oxygen. While they share the joint objective of harnessing sunlight for clean energy manufacture but these vary in their mechanisms and arrangements. Photo-electrolysis includes the utilization of a PEC cell or a photo-electrolysis cell, in which a semiconductor material absorbs light and enables the water-splitting reaction. The absorbed light creates electron-hole pairs in the semiconductor, which are utilized to drive the electrochemical reactions at the cathode and anode to yield hydrogen and oxygen. In photo-electrolysis, a solid-state semiconductor material is characteristically utilized as the photoactive element. The semiconductor absorbs light and starts the redox reactions at the electrode surfaces in interaction with water [19–21]. However, photochemical water splitting is a purely chemical process that does not use solid-state SCs. In this method, a photo-sensitizer (frequently a transition metal complex or a dye) absorbs light and creates excited-state electrons. These electrons are then utilized to drive the chemical reduction of water to generate hydrogen and oxygen. The photochemical water splitting characteristically takes place in a solution, where a photo-sensitizer is dissolved. It does not need a solid-state semiconductor. The excited electrons are utilized to start chemical reactions in the solution [22,23]. The schematic representation of water photo-electrolysis cell is shown in Figure 4.4.

Hydrogen evolution reaction (HER) is the main half-reaction to generate hydrogen at the cathode in water electrolysis that uses a two electron transfer phenomenon. The mechanism of this HER is extremely reliant on the experimental condition. There are three probable reaction stages for HER reaction in acidic medium.

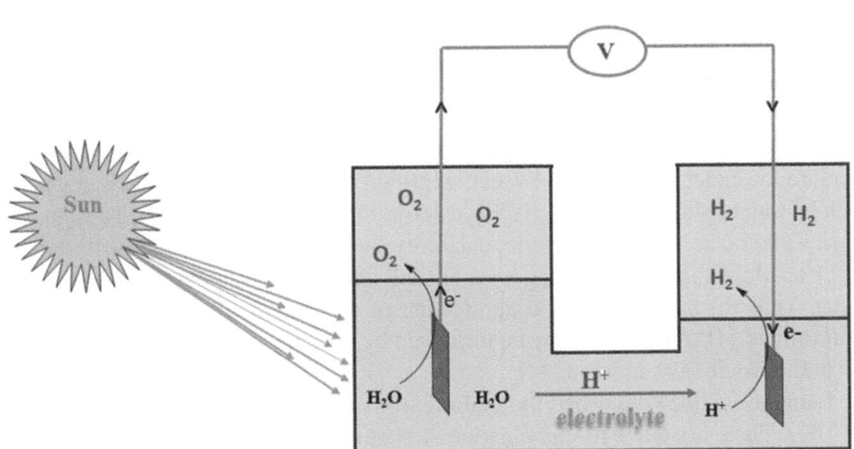

FIGURE 4.4 A schematic representation of water photo-electrolysis cell.

$$H^+ + e^- \rightarrow H_{ad} \qquad (4.3)$$

$$H^+ + e^- + H_{ad} \rightarrow H_2 \qquad (4.4)$$

$$2H_{ad} \rightarrow H_2 \qquad (4.5)$$

First step (Equation 4.3) is Volmer stage to generate adsorbed hydrogen. At that point, the HER occurred proceed by Heyrovsky stage (Equation 4.4) or the Tafel stage (Equation 4.5) or both to generate H_2 [24]. In case of alkali medium, HER reaction occurred via two probable reaction of Volmer stage (Equation 4.6) and Heyrovsky stage (Equation 4.7) [25] as given in the following equations, correspondingly:

$$H_2O + e^- \rightarrow OH^- + H_{ad} \qquad (4.6)$$

$$H_2O + e^- + H_{ad} \rightarrow OH^- + H_2 \qquad (4.7)$$

The simulation studies confirmed that HER activity depends on hydrogen adsorption (H_{ad}). The free energy and binding energy are very important parameter in HER. And by considering these aspects the electrolysis may be in alkaline, proton exchange membrane (PEM), and solid-oxide water electrolysis [26]. The solid-oxide water electrolysis shows high energy need due to high temperature required. PEM electrolysis water splitting is performed in acidic environment with benefits like low gas permeability and good proton conductivity [27]. This provides good energy efficacy and a considerable hydrogen generation rate [28]. Contrarily, the acidic medium restricts the oxygen evolution reaction (OER) on to noble metals-based photocatalysts [29] but the cost of OER cells decreased [30]. Thus, alkaline electrolysis cells water splitting has been studied for non-noble metals based photocatalysts in comparison to acidic medium [31]. Equally, HER in alkaline media is of two to three orders of scale in comparison to acidic medium [32]. Thus, a suitable plan is needed for inexpensive, good electro-catalysts, high catalytic activity, and various types of aqueous solutions.

In the development of photocatalysts the dissolution of noble metal based catalysts is the major problem while non-noble based catalysts cannot tolerate such conditions [33,34]. Therefore, there is a great demand for stable photocatalysts with good activity and performance. In this context non-noble metal-based photocatalysts have been prepared and used in basic medium but these felt structural and composition variations in OER [35,36]. For making the ideal catalysts, there is a great need to explore the exact mechanism of water splitting [37,38]. Therefore, hydrogen generation by water splitting is still a challenge due to the absence of a balanced design with varied photocatalysts having facile scalability, strong stability, and high affordability [39,40]. More than five decades have passed since the invention of water splitting using TiO_2 [6] but the practical application of photocatalytic splitting of water is still limited [41–44], which may be improved by varying various experiment factors including source of radiation [44].

Leung et al. [45] described the main reasons for low energy conversion efficacy of TiO_2. There are quick re-combination of photo-generated holes and electrons, and incapability to capture visible (greater than ~400 nm). Even then this material is the mot studied due to good resistance to corrosion [46]. Consequently, various materials have been developed and used for PEC cells [47]. Juodkazis et al. [48] used titanium oxide for hydrogen manufacture in photo-electrolysis. The authors described thermodynamics of energy conversion, i.e., light-to-electric, light-to-chemical, and electric-to-chemical. Seitz et al. [49] reported a photovoltaic-electrolysis system with the good solar to hydrogen (STH) efficacy utilizing two electrolyzers based on polymer membrane. The authors used these in series with of GaAs/InGaP/GaInNAsSb. The authors reported 30% STH in 48 hr.

Of course, alkaline water electrolyzers (AWEs), and PEM water electrolyzers, are the most extensively utilized electrolytic reactors in the world. The AWE is developed and made it simple to apply wide-scale utilization. Even then, but there are numerous concerns with AWEs, with high energy intake (more than 4 kWh/Nm^3). Alkaline electrolytes also interact with air carbon dioxide to produce insoluble carbonates (K_2CO_3), which transfer to the reactants. Though PEM water electrolyzers can attain smaller floor space yet these also experience a lot of problems such as costly equipment, high dependence on noble metals, and complex preparation procedures. Furthermore, there is a great need of reactors having solid oxide electrolyzers and anion exchange membrane. AEM electrolyzers have a long life but these are not fully developed.

4.5 DIRECT PHOTOCATALYTIC METHOD

The direct photocatalytic water splitting is a procedure that utilizes photocatalysis for the dissociation of water into hydrogen and oxygen by only light energy, water, and a catalyst(s) [50,51]. This process is assumed to be an artificial photo-synthesis that generates a solar fuel due to the reaction mimics natural photosynthetic oxygen generation and CO_2 fixation.

$$H_2O \rightarrow 2H_2 + O_2 \qquad (4.8)$$

The direct photocatalytic method of water splitting includes dispersing photocatalyst particles in water or depositing them on a substrate. The photocatalyst particles absorb light energy and utilize it to excite electrons, which then react with water molecules to generate hydrogen and oxygen. The most commonly used SC for water splitting are titanium dioxide (TiO_2), strontium titanate ($SrTiO_3$), and various other metal oxides ($BiVO_4$, Fe_2O_3, etc.). The photocatalytic activity of the catalyst is fixed by measuring the quantity of hydrogen and oxygen evolved from the water splitting. The absorption of sunlight through semiconductors surface starts the photolysis procedure [51,52]. Maeda et al. [53] used TaON photocatalyst with d0 electronic configuration for direct hydrogen photocatalysis. The variation of a less defective TaON (ZrO_2/TaON) with core/shell-structured RuO_x/Cr_2O_3 nanomaterials and colloidal IrO_2 was observed to be crucial to get the reaction. Molinari et al. [54] described high hydrogen generation rate of 0.41 mmol/h/g using M-30. It was observed that the amount of hydrogen produced was approximately 2.5 times than hydrogen generated by pure TiO_2 within 30 minutes. Jiang et al. [55] reported direct deionized water splitting for hydrogen production with 132.4 µmol/g/h evolution rate using piezo-photocatalysis. Furthermore, the authors reported 63.4 and 48.7 µmol/g/h evolution rates when using simulated and natural seawater.

It is significant to know the photo water-splitting mechanism before manipulating a good catalyst. Now, it is well known that the sun energy is taken up by semiconductors at their surfaces and utilized to split water into hydrogen and oxygen. About 4.3×10^2 J energy is reaching the earth from the sun every hour and about 2% is utilized. Therefore, there is a task to harvest this energy for our valuable drives. Fujishima and Honda [6] in 1973 reported the first hydrogen evolution using TiO_2 electrodes and UV (ultraviolet) radiation. The photocatalyst (semiconductor) absorbs photons from a light source, normally sunlight. This absorption of light excites electrons within the material, generating electron-hole pairs. The excited electrons (e^-) take parts in reduction reactions, while the holes (h^+) left behind participate in oxidation reactions. In the case of water splitting, the reduction reaction comprises the reduction of protons (H^+) to form hydrogen gas (H_2), and the oxidation reaction using the oxidation of water to produce oxygen gas (O_2).

The efficient separation of the generated electron-hole pairs is important to stop re-combination, where electrons and holes recombine before taking part in the desired redox reactions. This separation is frequently obtained by the design of the photocatalyst or by utilizing extra materials. The photocatalyst requires to be stable under the harsh situations of the water-splitting reaction. Besides, the complete efficiency of the process is effected by the wavelength of light absorbed, the band gap of the photocatalyst, and the surface area available for the reaction. Once the water-splitting reaction has occurred, the generated hydrogen and oxygen gases require to be collected and separated for further utilization. This may involve additional procedures like gas separation and purification.

The water-splitting reactions are given below.

$$H_2O \rightarrow 2H_2 + O_2 \tag{4.9}$$

The half-cell reactions are:

$$H_2O \;\; 2h^+ \rightarrow 2H_2 + \tfrac{1}{2}\,O_2 \rightarrow +2H^+ \tag{4.10}$$

$$2H^+ \;\; 2e^- \rightarrow 2H_2 \tag{4.11}$$

The water photo-splitting scheme is given in Figure 4.5.

The different photocatalysts have been synthesized and used for hydrogen production. As mentioned above the band gap is the key to hydrogen production. The main emphasis is given to the module on the band gap during the preparation of nanoparticles. Figure 4.6 shows the band gap of some important nanoparticles.

During the write-up of this book some photocatalysts were found good in their performance. These are briefed in Table 4.1.

The direct photocatalytic water splitting is an exciting problem due to uphill thermodynamic reaction. The light absorption results in the development of energetic charge-separated stages in both molecular donor-acceptor schemes and

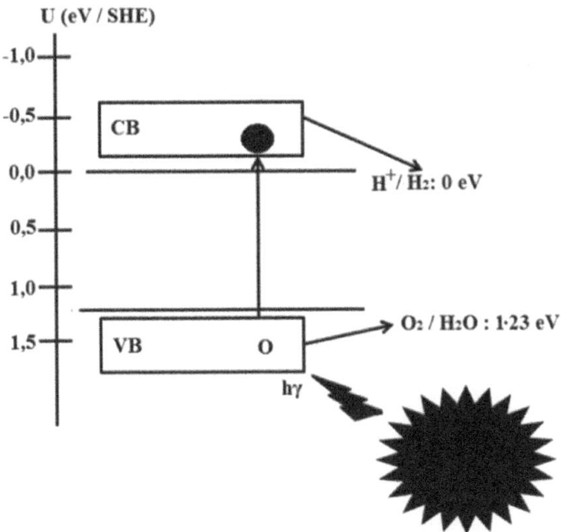

FIGURE 4.5 The mechanism of water splitting [reproduced with permission from ref. 56].

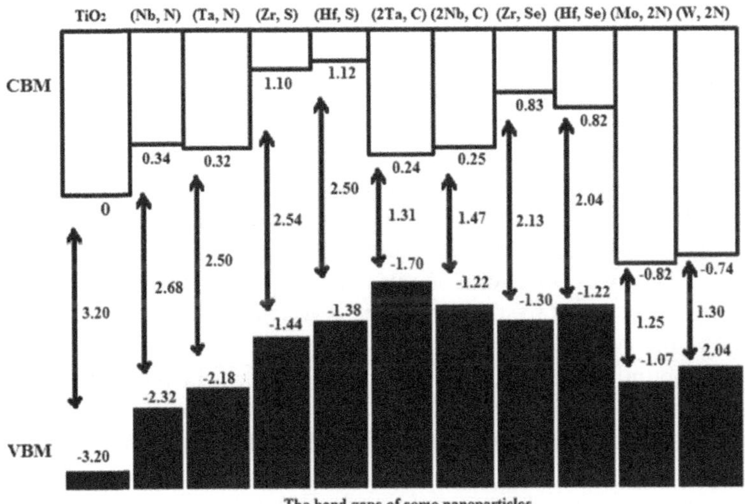

FIGURE 4.6 The band gaps of some photocatalysts [reproduced with permission from ref. 56].

semiconductor particles. But the energy encouraging charge re-combination reactions incline to be much quicker than slow multi-electron routes of water oxidation and reduction. Therefore, visible light water splitting is gaining importance achieved in water splitting. The researchers continue to design new materials and improve the efficacy of photocatalysts to make direct photocatalytic water

TABLE 4.1

Good performance photocatalysts and co-catalysts

Photocatalysts	Co-catalysts	Efficiency	Radiation	Reference
Titanium oxide	Pt/RuO$_2$	QE: 30 ± 10% (QE)	310 nm	[57]
Titanium-lanthanum oxide with barium	Varied nitric oxides	35% (QE)	<360 nm	[58]
Strontium-niobium oxide	Nickel	23% (QE)	<300 nm	[59]
Sodium-tellurium oxide with lanthanum	Nickel oxide	56% (AQE)	270 nm	[60]
Galium oxide with zinc	Rh$_{2-y}$Cr$_y$O$_3$	71% (AQY)	254 nm	[61]
GaN:Mg/InGaN:Mg	Chromium oxide with ruthenium	12.3% (AQE)	400–475 nm	[62]
Quantum dots of tricyanoamine	-	16% (AQE)	420 nm	[63]

splitting a good and sustainable approach for hydrogen production. For effective photocatalysts, these need to meet multiple important criteria, like inexpensiveness, stability, and good quantum efficacy. Briefly, light energy, water, and a catalyst (or catalysts) are the only materials required for the direct photocatalysis of water, which utilizes photocatalysis to split water into hydrogen and oxygen. Nonetheless, obstacles need to be addressed before this procedure may be used on a large industrial basis.

It is the time to analyze the performance of all the photocatalytic methods discussed above. Of course, a considerable progress has been done in solar-driven water-splitting (photo-thermal, PEC, photo-electrolysis, and direct photocatalytic) methods using a variety of photocatalysts in nanoparticles form, specially TiO$_2$, but no practical method has yet been identified for hydrogen production at a large scale economically [64]. There are several challenges to overcome scenically and technically. The hydrogen evolution rates depend on the experimental setup, like type of catalysts, co-catalyst, light source wavelength, light intensity and duration, size of the potential size, and water conductivity. Therefore, there is a great demand to optimize the experimental conditions for achieving good amount of hydrogen production. The calculation of apparent quantum yield (AQY) in place of only gas evolution should be used [65,66]. Most of the photocatalysts have suitable band gaps that can absorb UV radiation and such materials are not useful in the solar spectrum. Therefore, there is a great need to optimize the band gaps of the photocatalysts. The doping agent types and concentration can be optimized by docking studies. As per Parzinger et al. [67] and Li et al. [68] the combination of MoS$_2$ and WS$_2$ with TiO$_2$ could be better photocatalysts in the future. It is also important to synchronize the computer modeling results with experimental ones to optimize the performance of the photocatalysts. There is a lack of scalable systems for economic hydrogen production at large scale. Zhao et al. [69] reported 1.8% solar to hydrogen efficiency at scalar level. This data clearly indicates the future need to improve all the aspects of photocatalytic water splitting to compete with other methods.

4.6 RADIOLYSIS METHOD

Water radiolysis is a process that uses ionizing radiation to break down water (H_2O) into its component elements, hydrogen (H_2) and oxygen (O_2). This process involves the energy deposited by high-energy radiation, typically ionizing radiation like gamma rays, X-rays, or high-energy particles (e.g., electrons, protons, neutrons), breaking chemical bonds in water molecules. Many uses for water radiolysis have been investigated, including the production of hydrogen for nuclear and space exploration.

Debierne was the first researcher to propose that the production of the H atom and the OH radical is the mechanism by which water is radiolyzed [70]. Since then, other workers have tried to apply this method to a variety of items, most notably hydrogen for green energy. According to Yamada et al. [71], hydrogen can be produced by radiolysis of water in 0.4 M H_2SO_4 with alumina. By increasing the absorbed dose rate in the range of 1–5 kGy/h, the scientists noticed a rise in the amount of hydrogen. Crumière et al.'s study [72] examined how cyclotron and gamma radiation affect the synthesis of hydrogen. Linear energy transfer values ranged from 0.23 to 151.5 keV/μm. ZrO_2 nanopowders for water radiolysis to create green hydrogen were reported by Southwortha et al. [73]. The authors optimized hydrogen production and reported good quantity of hydrogen produced. McGrady et al. [74] reported the water radiolysis on ZrO_2 nanoparticles' surface using gamma radiation. The authors reported increased yields of H_2 and O_2 on increasing the nanoparticles' surface area. Kumar et al. [75] described a Monte Carlo simulation of water radiolysis. The authors told that gamma water radiolysis produced excited hydrogen atoms, giving to hydrogen production.

Nowadays, the maximum work in this direction is being done by two groups of India and Azerbaijan. These authors utilized various photocatalysts to create hydrogen from water splitting by gamma radiation attack. The most significant photocatalysts used are alumina [76,77], zirconium [78,79], and niobium composites [80]. These workers utilized a mixture of pure water, seawater, and water with organic solvent with above-mentioned photocatalysis. These workers also did radiolysis in thermal, radiation, and radiation-thermal approaches. Generally, the yields of hydrogen were radiation-thermal > radiation > thermal modes. The effect of temperature, amount of catalysts , radiation time, the particle size of the catalysts, and dose were studied. The nano-sized catalysts were found to be active for creating good quantity of hydrogen. Generally, in their labors these workers reported a good quantity of hydrogen manufacture. Currently, these authors reported the modeling of hydrogen manufacture using water and hexane mixture. The authors stated numerous rates of hydrogen formation procedures (W) for H_2, CH_4, C_2H_6, C_3H_8, C_4H_{10}, and C_5H_{12} as 3, 0.12, 0.56, 0.57, 0.3, and 0.06×10^{-14} molecule/s. Monte Carlo modeling documented the kinetics of the radicals and radiolysis products. It was well-known that the reactions occurred in spurs regarding the hydrated electron eaq, $H^{\bullet}$, H^+, $OH^{\bullet}$, H_2, OH^-, and H_2O_2, and various reactions including these particles was 10. Similar workers presented a motivating review article on radiation role on water splitting. The workers described numerous radiations for water splitting with diverse catalysis [81]. The radiation-chemical yields of seawater without a catalyst, with Zr1%Nb (Zr = 99%, Nb = 1%) and alone Zr, were G(H_2) = 0.81, 437.4, and 307.1 molecules/100 eV,

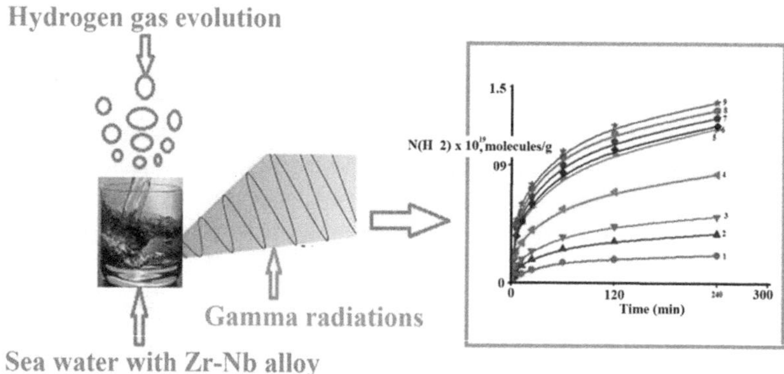

FIGURE 4.7 Kinetics of accumulation of molecular hydrogen [reproduced with permission from ref. 80].

correspondingly. The radiation–thermal water decay augmented in γ-radiation of Zr1%Nb + SW system with an elevating temperature. At 1,273 K, it triumphs over radiation procedures. In radiation and heat radiation heterogeneous processes in Zr1%Nb + SW, the manufacture of surface energetic sites and secondary electrons enhanced the growth of molecular hydrogen and oxidation of Zr1%Nb. Thermal and thermal radiation methods produced metal phase to accumulate thermal surface energetic sites for water break and oxidation of Zr1%Nb beginning at 573K as shown in Figure 4.7 [80].

Water radiolysis has been utilized in nuclear reactors to produce hydrogen for many applications, comprising cool and as a potential energy carrier. Still, radiolysis is not effective in manufacturing hydrogen at a bulky scale. It requires the optimization of the radiolysis procedures. The most significant parameters that require to be optimized are source of radiation, dose of radiation, and by-products of radiolysis. There is an abundant requirement for high-energy radiation sources such as particle accelerators or nuclear reactors. The controlling cautious doses of both catalysts and radiation is of utmost importance. There is also a big demand to advance catalysis, preferably nanoparticles for a good quantity of hydrogen manufacture. There is also a big need to discrete green hydrogen from other by-products of radiolysis $(H_2, HO\cdot, H\cdot, H_2O\cdot, H_3O^+, OH^-, H_2O_2,$ etc.) as many products are generated in this process in water splitting. The minimalizing and handling the manufacture of these by-products are crucial. The safety alarms are also significant as high-energy radiation is used. The good class of hydrogen can be manufactured by using good quality of water. Consequently, water quality is also of great reputation in this procedure. Some other papers relating to direct water splitting in UV and visible radiations are summarized in Table 4.2.

4.7 MECHANISM OF WATER RADIOLYSIS

The mechanism of water radiolysis contains the breakdown of water (H_2O) into hydrogen and oxygen and formation of numerous radicals, due to the interactions of

TABLE 4.2

Hydrogen generation from water splitting by nuclear radiations

Sl. No.	Photo-materials	Dose of γ-rays	H_2 generated	Refs.
1.	Alpha-, gamma-, and delta alumina	0.40 Gy/s	4.50–7.0×10^{18} molecules/g/30h	[76]
2.	Alumina	1.40 Gy/s	5.96 molecules/100 eV	[77]
3.	Zirconium oxide	0.40 Gy/s	8.1–10.1×10^{18} molecules/g.6h	[78]
4.	Zirconium oxide	0.11 Gy/s	10.1–20.1×10^{18} molecules/g.6h	[79]
5.	Zirconium-niobium composite	0.40 Gy/s	1.80×10^{18} molecules/g.5h	[80]
6.	Chromium and iron oxides	8.0 kGy/h	18.20 molecules/g	[82]
7.	Silicon oxide	28.4 kGy/h	2.13 molecules/100 eV	[83]
8.	Waste materials with plutonium and uranium oxides	*10–20 MGy	0.56 molecules/g	[84]
9.	Plutonium oxide	-	9.98×10^{-6}–2.06×10^{-3} cm^3/g.day	[85]

*: alpha rays

ionizing radiation with water molecules. The mechanism of water radiolysis is sound recognized as numerous workers tried to explain it [86–89]. The whole reaction of water hydrolysis is given below.

$$H_2O + \text{Radiation} \rightarrow e^- + H_2, \; HO, \; OH^-, \; H.,$$

$$H_2O^{\cdot}, \; HO_2^{\cdot}, \; H_2O_2, \; H_3O^+, \text{etc.} \tag{4.12}$$

Essentially, this reaction is done via physical, physico-chemical, and chemical phases. The physical phase is achieved in about 1.0 femto seconds just after radiation interactions with catalyst and comprises of energy confession followed by quick relaxation occurrence. This delivers to the formation of excited water molecules (H_2O^*), ionized water molecules (H_2O^+), and sub-excitation electrons (e^-). The physico-chemical phase (10^{-15}–10^{-12} s) includes many procedures to happen, with dissociative relaxation, ion-molecule reaction, hole diffusion, thermalization of sub-excitation electrons (solvation of electrons), auto-ionization of excited states, etc. The reactions happening in this stage are shown below.

$$H_2O^+ + H_2O \rightarrow H_3O^+ + HO^{\cdot} \tag{4.13}$$

$$H_2O^* + HO_2 \rightarrow HO^{\cdot} + H \tag{4.14}$$

$$e^- \rightarrow e^-_{eq} \tag{4.15}$$

The third chemical phase (10^{-12}–10^{-6} seconds) comprises the interactions of numerous species trailed by their dispersal. Therefore, these species react with one another with

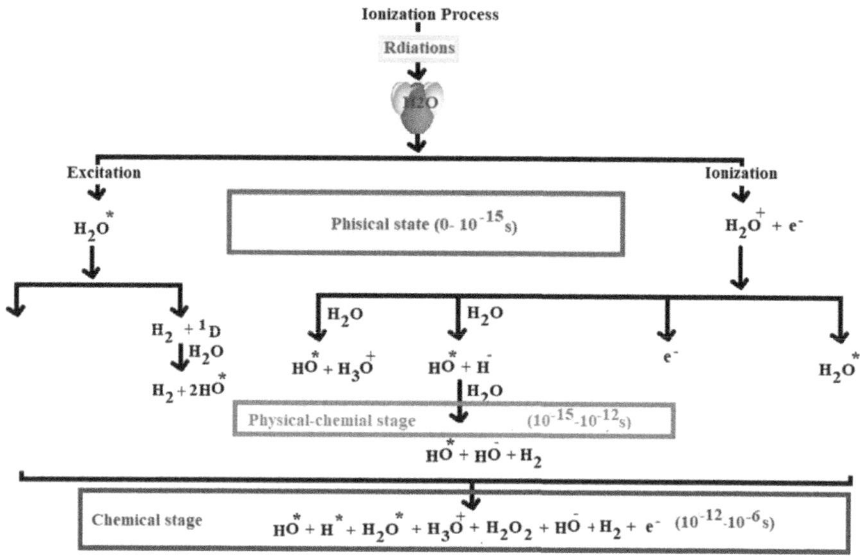

FIGURE 4.8 Mechanism of water radiolysis [reproduced with permission from ref. 89].

adjacent molecules. The track of the particles increases since of the radicals' diffusion and their following chemical reactions. These three phases are given in Figure 4.8 [89].

The radiolysis procedure of water in heterogeneous systems is powerfully effected by the solid-liquid border, and all three (physical, physico-chemical, and chemical) phases of the procedure in those schemes vary from deionized water to other liquid solutions. It should be observed that, when ionizing rays go via the substance, energy transfer happens by interactions in molecules or atoms. This leads to the development of electron-excitation, electron-ion pairs, active intermediate particles, and other species. The catalysts play a vital part in all these interactions and disintegrations. The holes and electrons created in the catalyst owing to radiation are caught by volume defects. The holes travel to the catalysts and are taken over by the surface and .O group and form H$^+$ are formed aa shown below:

$$h^+ + -OH \rightarrow -O^* + H^+ \tag{4.16}$$

Later, H$^+$ arrests electrons (e$^-$) on the surface and an H* is produced as shown below:

$$H^+ + e_t^- \rightarrow H^* \tag{4.17}$$

Merely a few workers have paid consideration to the electrons that are generated by the influence of ionizing rays on heterogeneous systems, which play a main role in the creation of hydrogen. Irrespective of the mechanism of gaining, two hydrogen atoms associate together to form the hydrogen molecule as shown below:

$$H^* + H^* \rightarrow H_2 \tag{4.18}$$

4.8 EFFECT OF DIFFERENT TYPES OF PHOTOCATALYSTS

This book is dedicated to hydrogen production by water splitting in the presence of photocatalysts. During the last decade, many photocatalysts of nano dimensions have been prepared and utilized for this purpose. Therefore, this sub-heading deals with the effect of various nano-photocatalysts on hydrogen production. Nanomaterials have shown exciting applications in different areas of science and technology because of their special physical and chemical properties compared with bulk materials. They can be synthesized using bottom-up or top-down approaches, and, with recent advances, it is possible to have better control of size, shape, and structure by adopting a suitable synthesis approach and manipulating the synthesis conditions. Depending on their type, size, shape, and structure, nanomaterials demonstrate a range of unique electrical, optical, mechanical, magnetic, and antimicrobial properties, which have led to many interesting technological applications.

Specific surface area, particle size, crystallinity, crystalline phase, and shape are the most significant characteristics of nanoparticles influencing hydrogen generation [90]. Wide band-gap SCs can be doped with metals or nonmetals to reduce their band gap and to apply catalysts to enhance charge separation. This gives the SC's surface active spots for water splitting. By improving charge separation, multicomponent heterojunctions made of several SC nanomaterials provide an efficient way to prolong solar absorption and lengthen the lifetime of charge carriers. Water splitting is always improved by a large surface area and tiny particle size. When nano-sized TiO_2 is combined with appropriate narrow band-gap SCs, nanocomposite photocatalysts are created that demonstrate enhanced hydrogen generation from the visible light spectrum. Due to its appropriate electrical band structure and very tiny band gap, CdS has been extensively researched for hydrogen production. Effective nanocomposites for hydrogen production can be created by combining CdS with other SCs to enhance CdS's charge separation and stability. Among the greatest photocatalysts for producing hydrogen under visible light are those based on CdS. The creation of hydrogen is controlled by variables like composition, the interface between SCs, and the shape of each component. Furthermore, the structure of the hybrid has a sensitive and significant impact on the charge transfer in multi-component photocatalysts. Therefore, to maximize the electron transport throughout the photocatalyst, the relative positions of the SC and cocatalysts must be regulated [91]. Furthermore, it was discovered that the production of hydrogen increased linearly with an increase in the metal-work function. As the work function difference between TiO_2 and the metal decreases, so does the rate of charge recombination.

4.9 EFFECT OF DIFFERENT TYPES OF RADIATIONS

Due to its ability to create holes and electrons on the surface of solid materials, radiation plays a crucial role in the splitting of water. In photo-materials, electrons and holes are present prior to radiation irradiation. Different radiation types can have varying effects on various processes. Photon absorption is the means by which SC materials produce electron-hole pairs for PEC water splitting. The UV or visible light is necessary for this procedure and can increase the effectiveness of water splitting.

Indeed, enhanced PEC performance could be achieved by optimizing particular wavelength bands. UV light is one type of non-ionizing radiation that can be utilized to increase electrolysis efficiency. In order to improve the overall water-splitting process, UV radiation can encourage the dissociation of water molecules at the electrode surface. High-energy ionizing radiation has the ability to harm the electrodes and electrolyte, two components employed in the electrolysis process. It could result in the electrolysis cell's performance deteriorating or its efficiency declining. However, the radiation dose and the particular materials utilized determine how much damage is done. The SC components used in PEC cells may be impacted by ionizing radiation. It might cause flaws in the substance or alter its electrical characteristics, which could have an impact on how well water-splitting works. Studies have indicated that some materials may be more resistant than others to harm caused by radiation.

No experimental paper was found to report the effect of various radiations on water splitting for hydrogen generation. A review article is already written by the authors of this book and it was used to compare the application of the different radiations for hydrogen generation [92]. From this article, it was observed that the hydrogen production was 6,500 and 19,800 μmol using visible UV radiation on the surface of the titanium catalyst. Another observation was that 853 μmol hydrogen was produced on a zirconium catalyst using UV radiation while the value of hydrogen generation was 3 to 8 molecules using gamma radiation. Besides, other observations were also made in this review article and it was found that the hydrogen production was directly proportional to the extent of the energy of the radiations. In a nutshell, hydrogen production is effected by the materials employed, the particular technique used, and the type of radiation. Non-ionizing radiation, such UV or visible light, can sometimes be advantageous by speeding up the water-splitting process, whereas ionizing radiation has the potential to harm materials and affect performance. The effects of radiation on these processes are still being investigated, and potential solutions are being looked into.

4.10 EFFECT OF DIFFERENT TIMES OF RADIATION IRRADIATION

Radiation irradiation, when used to produce hydrogen by water splitting, is a complicated process that relies on a number of variables, such as the kind, intensity, duration, and materials utilized. Generally speaking, there are two ways to divide water: photolysis and radiolysis. The wavelength and intensity of the light employed determine how effective photolysis is. Shorter wavelengths (UV light, for example) have more energy and can split water more efficiently. The total hydrogen output during photolysis can vary depending on the duration of irradiation. Long-term exposure to light may increase the generation of hydrogen up to a degree, at which time the system may become saturated and the rate of production may plateau. Radiation's impact on water splitting varies with exposure time and dose. In summary, a variety of parameters, such as the type of radiation, its intensity, and the materials and catalysts employed in the process, affect how varied durations of radiation irradiation affect the creation of hydrogen by water splitting. In order to produce hydrogen using water-splitting techniques with greater efficiency and yield, these factors must be optimized. It is also critical to note that extended exposure to high radiation doses

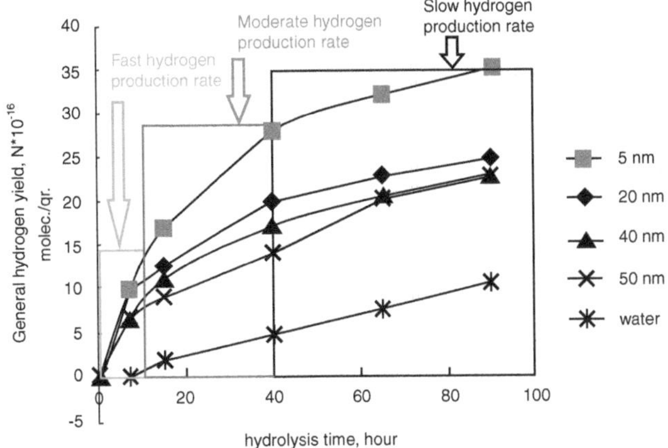

FIGURE 4.9 Effects of time on hydrogen generation. Other conditions: m_{water}/m_{cat} = 100:1, m_{cat} = 0.01 g, T = 300 K, pressure = 1.04 kGy/h [reproduced with permission from ref. 77].

can harm process materials and result in the production of undesirable byproducts, particularly in the case of gamma and X-ray radiations.

Not much work has been carried out in this direction. However, some papers describe the effect of radiation duration on hydrogen production. Ali et al. [81] reported the hydrogen production in the period of 1–5 hr. The radiation used gamma with α-, δ-, and γ-Al$_2$O$_3$. In the case of thermal catalytic decomposition, the values of hydrogen accumulations at 673 K using 1, 5, and 10 mg/cm³ increased sharply to 0.10, 0.20, and 0.25 at 5 hr time. Similar results were obtained with zirconium catalyst [76]. The values of hydrogen increased from 50 to 150 minutes with maximum hydrogen production of 1.75×10^{16} molecules. Further increase in irradiation time could not lead to hydrogen production. Zirconium and its niobium alloy were used under gamma radiation for water splitting [93]. The authors allowed gamma radiation irradiation from 30 to 600 minutes. The authors reported seawater splitting under thermal, radiation, and radiation–thermal processes. The maximum hydrogen yields ranged from 120 to 150 minutes. The same group carried out an experiment for 100 hr with gamma radiation and alumina catalyst [78]. The maximum yield of hydrogen was at 60 hr. The authors reported three zones of hydrogen production, i.e., fast, moderate, and slow regions (Figure 4.9).

4.11 EFFECT OF DIFFERENT TEMPERATURES

Temperature has a major impact on the process of water splitting, which produces hydrogen. The rate of the reaction and the process's energy efficiency are both impacted by temperature. Because the water-splitting reaction is an endothermic process—that is, it needs energy input to break the bonds in water molecules—it proceeds more slowly at low temperatures. Even at low temperatures, the electrolysis process can still take place, although it will need less energy. In comparison to higher temperatures, more electrical energy is needed to create the same amount of hydrogen. Because of its decreased efficiency, low-temperature electrolysis is frequently less feasible for large-scale hydrogen production. As the temperature climbs within this range, the water-splitting reaction accelerates. Raising the temperature facilitates the breakdown of water molecules and the release of hydrogen and oxygen by lowering the activation energy needed for the reaction. In comparison to low temperatures, electrolysis is more effective at moderate temperatures. A specific amount of hydrogen can be produced using less electrical energy.

The efficiency of water splitting increases with temperature, both in terms of reaction rate and energy usage. Because steam is frequently used as the electrolyte in high-temperature electrolysis, the process is sometimes known as "steam electrolysis" or "thermochemical electrolysis." For the creation of hydrogen, high-temperature electrolysis may be more energy-efficient, particularly if surplus heat can be collected and put to better use. It is crucial to remember that the precise temperature range needed to produce the most hydrogen can change based on the particular catalysts and materials employed in the electrolysis process. Certain cutting-edge materials and catalysts are appropriate for specific applications because of their ability to function well at lower temperatures.

The process that splits the water is an endothermic reaction. As a result, we may anticipate that, when the temperature rises, the water-splitting equilibrium will alter to create more hydrogen. Thermochemical water splitting produces hydrogen and oxygen from water by heating it to high degrees using concentrated solar power, nuclear power waste heat, or chemical reactions. This is a long-term technological route that might emit little or no greenhouse gases. In summary, temperature affects hydrogen production through water splitting significantly. Although the precise temperature range may vary depending on the tools and materials used in the process, higher temperatures often result in faster and more energy-efficient electrolysis. The researchers continue to work on developing more effective and cost-effective electrolysis approaches for hydrogen production at several temperature ranges to support clean energy applications.

The important factors governed by temperature are chemical equilibrium, rate-determining steps, and plasmon resonance. Besides the catalyst properties such as band gaps, onset potential, intrinsic carriers, and localized surface are also affected [77]. The change in these properties affects the hydrogen production. Not much work has been carried out on hydrogen production by varying the temperature. However, Imran Ali and his colleagues did some experiments on this issue. Ali et al. [94] reported hydrogen production at 673, 773, 873, and 973 K using α-, δ-, and γ-Al_2O_3 as catalysts. The hydrogen production was found to be temperature-dependent. The values of hydrogen amounts were 10, 16, 20, and 23 $N(H_2) \times 10^{18}$ molecules/g at 673, 773, 873, and 973 K. Similar results were obtained at 373, 473, 573, and 673 K at nano-ZrO_2 surface [95]. In another study, the same group [31] reported an increasing hydrogen product at high temperatures when using a water-n-hexane-ZrO_2 system. The temperature used were 373, 473, and 573 K and the maximum hydrogen was reported at 573 K temperature. In another study, a wide range of temperatures, i.e., 473, 573, 673, 773, 873, 973, 1,073, 1,173, and 1,273 K, were used [93]. The authors reported an increase in hydrogen production at elevated temperatures. Briefly, hydrogen production may be increased at high temperatures by harnessing thermal energy in terms of reaction kinetics, as well as material selection. This will not only provide a synergistic effect to improve conversion efficiency but can also be used as a method to compensate for losses occurring in a typical up-scaled PEC device.

4.12 EFFECT OF DIFFERENT AMOUNTS OF WATER AND ORGANIC SOLVENTS

Depending on the particular solvent and the circumstances of the reaction, organic solvents can have varying effects on water splitting for the production of hydrogen. The efficiency and safety of the reaction might be affected by the solvent selected for this procedure, which usually takes place in an aqueous solution. Depending on the particular solvents and the reaction circumstances, using a combination of organic solvents and water to generate hydrogen might have benefits and drawbacks. These combinations are frequently employed to alter the electrolyte's characteristics in an effort to increase the process's water-splitting efficiency. The aqueous solution's electrical conductivity can be improved by adding organic solvents. This may result in ions being transported more effectively, which would lower the energy needed for the water-splitting reaction. It is also possible to customize the electrolyte's characteristics by selecting organic solvents. They can be chosen to have low freezing temperatures or high boiling points, for instance, which might be useful under specific operating circumstances. Certain organic solvents might be more stable than pure water under extreme circumstances or at high temperatures. The electrolysis cell's lifespan may be prolonged in this way. The mixture's organic solvent selection may have an impact on the water-splitting process's selectivity, enabling the favored synthesis of oxygen or hydrogen.

The usage of organic solvents has certain disadvantages despite its benefits. Many organic solvents have the potential to burn and be dangerous. Working with such mixes requires the adoption of appropriate safety precautions, particularly when electrical equipment is present. Certain electrolytes can lose some of their solubility over time, which results in inadequate hydrogen generation. The system becomes more difficult when solvent mixes are used, since the mixture's composition must be carefully chosen and controlled. For some applications, the presence of organic solvents in the electrolyte may cause contaminants to enter the hydrogen gas generated, so affecting its purity. The environmental impact of organic solvents should be taken into account while selecting them, particularly if they are volatile or have unfavorable impacts on the environment.

In summary, increasing the performance and efficiency of the water-splitting process can be achieved by combining a blend of organic solvents with water in the hydrogen-generating process. But it necessitates giving considerable thought to the particular solvents, safety precautions, and the process's intended results. To improve hydrogen generation outcomes while addressing safety and environmental concerns, researchers are still investigating various solvent systems and electrolyte formulations.

Some papers describe the effect of organic solvents on hydrogen production in the water-splitting phenomenon. Yamaguchi et al. [96] studied a mixture of methanol and water to produce hydrogen using iridium complex catalysts. The authors reported a good amount of hydrogen while removing the hurdle of the problems about homogeneous catalysts. The authors used an iridium catalyst for hydrogen production in a methanol and water mixture. The authors reported a decreased amount of hydrogen with increasing amounts of methanol. This was due to a decrease in the activity of the catalyst. Ali et al. [78] used a water and n-hexane mixture to produce hydrogen on zirconium oxide surface. Gamma radiation was used as the source of energy. The experiments were carried out with hexane-water of 0.66 g:1.0 g for liquid and 0.0051 g:0.0015 g in a vapor phase. The amount of hydrogen produced varied with the different concentrations of hexane and water.

REFERENCES

1. W.K. Fan, M. Tahir, Recent developments in photothermal reactors with understanding on the role of light/heat for CO2 hydrogenation to fuels: A review, *Chem. Eng. J.*, 427, 131617 (2022).
2. P. Fan, Y. He, J. Pan, N. Sun, Q. Zhang, C. Gu, K. Chen, W. Yin, L. Wang, Recent advances in photothermal effects for hydrogen evolution, *Chinese Chem. Lett.*, 35, 108513 (2023).
3. C. Muhich, B. Ehrhart, I. Alshankiti, B. Ward, C. Musgrave, A. Weimer, A review & perspective of efficient hydrogen generation via solar thermal water splitting: A review & perspective of efficient hydrogen generation, *Wiley Interdiscip. Rev. Energy Environ.*, 5, 261–287 (2015).
4. R. Abe, Recent progress on photocatalytic & photoelectrochemical water splitting under visible light irradiation, *J. Photochem. Photobiol. C Photochem. Rev.*, 11, 179–209 (2010).
5. A. Kudo, Y. Miseki, Heterogeneous photocatalyst materials for water splitting, *Chem. Soc. Rev.*, 38, 253–278 (2009).
6. A. Fujishima, K. Honda, Electrochemical photolysis of water at a semiconductor electrode, *Nature*, 238, 37–38 (1972).
7. J. Brillet, J.H. Yum, M. Cornuz, T. Hisatomi, R Solarska, J Augustynski, M. Graetzel, K. Sivula, Highly efficient water splitting by a dual-absorber tandem cell, *Nat. Photonics*, 6, 824–828 (2012).
8. B.D. Alexander, P.J. Kulesza, I. Rutkowska, R. Solarska, J. Augustynski, Metal oxide photoanodes for solar hydrogen production, *J. Mater. Chem.*, 18, 2298–2303 (2008).
9. C. Du, X. Yang, M.T. Mayer, H. Hoyt, J. Xie, G. McMahon, G. Bischoping, D. Wang, Hematite-based water splitting with low turn-on voltages, *Angew. Chem. Int. Ed.*, 52, 12692–12695 (2013).
10. J.W. Jang, C. Du, Y. Ye, Y. Lin, X. Yao, J. Thorne, E. Liu, G. McMahon, J. Zhu, A. Javey, J. Guo, D. Wang, Enabling unassisted solar water splitting by iron oxide & silicon, *Nat. Commun.*, 6, 7447 (2015).
11. A. Malathi, J. Madhavan, M. Ashokkumar, P. Arunachalam, A review on BiVO4 photocatalyst: Activity enhancement methods for solar photocatalytic applications, *Appl. Catal. A: Gen.*, 555, 47–74 (2018).
12. A. Kudo, K. Omori, H. Kato, A novel aqueous process for preparation of crystal form-controlled & highly crystalline BiVO4 powder from layered vanadates at room temperature & its photocatalytic & photophysical properties, *J. Am. Chem. Soc.*, 121, 11459–11467 (1999).
13. A. Malathi, V. Vasanthakumar, P. Arunachalam, J. Madhavan, M.A. Ghanem, A low cost additive-free facile synthesis of BiFeWO6/BiVO4 nanocomposite with enhanced visible-light induced photocatalytic activity, *J. Colloid Interface Sci.*, 506, 553–563 (2017).

14. Q. Mi, A. Zhanaidarova, B.S. Brunschwig, H.B. Gray, N.S.A. Lewis, A quantitative assessment of the competition between water & anion oxidation at WO3 photoanodes in acidic aqueous electrolytes, *Energy Environ. Sci.*, 5, 5694–5700 (2012).
15. S.Q. Yu, Y.H. Ling, J. Zhang, F. Qin, Z.J. Zhang, Efficient photoelectrochemical water splitting & impedance analysis of WO32x nanoflake electrodes, *Int. J. Hydrogen Energy*, 42, 20879–20887 (2017).
16. W. Feng, L. Lin, H. Li, B. Chi, J. Pu, J. Li, Hydrogenated TiO2/ZnO heterojunction nanorod arrays with enhanced performance for photoelectrochemical water splitting, *Int. J. Hydrogen Energy*, 42, 3938–3946 (2017).
17. P. Arunachalam, K. Nagai, M.S. Amer, M.A. Ghanem, R.J. Ramalingam, A.M. Al-Mayouf, Recent developments in the use of heterogeneous semiconductor photocatalyst based materials for a visible-light-induced water-splitting system, A brief review, *Catalysts*, 11, 160 (2021).
18. K. Wenderich, W. Kwak, A. Grimm, G.J. Kramer, G. Mul, B. Mei, Industrial feasibility of anodic hydrogen peroxide production through photo-electrochemical water splitting: A techno-economic analysis, *Sustain. Energy Fuels*, 4, 3143 (2020).
19. J. Jia, L.C. Seitz, J.D. Benck, Y. Huo, Y. Chen, J.W.D. Ng, T. Bilir, J.S. Harris, T.F. Jaramillo, Solar water splitting by photovoltaic-electrolysis with a solar-to-hydrogen efficiency over 30%, *Nat. Commun.*, 7, 13237 (2016).
20. H. Eidsvåg, S. Bentouba, P. Vajeeston, S. Yohi, D. Velauthapillai, TiO2 as a photocatalyst for water splitting—An experimental & theoretical review, *Molecules*, 26, 1687 (2021).
21. S. Wang, A.L.C.J. Zhong, Hydrogen production from water electrolysis: Role of catalysts, *Nano Converg.*, 8, 1–23 (2021).
22. A.A. Ismail, D.W. Bahnemann, Photochemical splitting of water for hydrogen production by photocatalysis: A review, *Sol. Energy Mater. Sol. Cells*, 128, 85–101 (2014).
23. I.U. Hassan, G.A. Naikoo, H.Salim, T.Awan, M.A. Tabook, M.Z. Pedram, M. Mustaqeem, A. Sohani, S. Hoseinzadeh, T.A. Saleh, Advances in photochemical splitting of seawater over semiconductor nano-catalysts for hydrogen production: A critical review, *J. Ind. Eng. Chem.*, 121, 1–14 (2023).
24. E. Skúlason, G.S. Karlberg, J. Rossmeisl, T. Bligaard, J. Greeley, H. Jónsson, J.K. Nørskov, Density functional theory calculations for the hydrogen evolution reaction in an electrochemical double layer on the Pt(111)
25. N.M. Marković, S.T. Sarraf, H.A. Gasteiger, P.N. Ross, Hydrogen electrochemistry on platinum low-index single-crystal surfaces in alkaline solution, J. Chem. Soc. Faraday Trans. 92(20), 3719–3725 (1996).
26. F.M. Sapountzi, J.M. Gracia, C.J. (Kees-Jan) Weststrate, H.O.A. Fredriksson, J.W. (Hans), Electrocatalysts for the generation of hydrogen, oxygen and synthesis gas, Prog. Energy Combust. Sci., 58, 1–35 (2017).
27. S. Wang, A. Lu, C.J. Zhong, Hydrogen production from water electrolysis: role of catalysts, Nano Converg., 8, 4 (2021).
28. C. Hu, L. Zhang, J. Gong, Recent progress made in the mechanism comprehension and design of electrocatalysts for alkaline water splitting, Energy Environ. Sci., 12, 2620–2645 (2019).
29. H.Y. Qu, X. He, Y. Wang, S. Hou, Electrocatalysis for the Oxygen Evolution Reaction in Acidic Media: Progress and Challenges, Appl. Sci., 11 432 (2021).
30. K. Sardar, E. Petrucco, C.I. Hiley, J.D.B. Sharman, P.P. Wells, A.E. Russell, R.J. Kashtiban, J. Sloan, R.I. Walton, Water-splitting electrocatalysis in acid conditions using ruthenate-iridate pyrochlores, Angew. Chem., 126, 11140–11144 (2014).
31. J. Brauns, T. Turek, Alkaline water electrolysis powered by renewable energy: A review, *Processes*, 8, 248 (2020).

32. D. Strmcnik, M. Uchimura, C. Wang, R. Subbaraman, N. Danilovic, D. van der Vliet, A.P. Paulikas, V.R. Stamenkovic, N.M. Markovic, Improving the hydrogen oxidation reaction rate by promotion of hydroxyl adsorption, Nat. Chem., 5, 300–306 (2013).
33. A. Vazhayil, L. Vazhayal, J. Thomas, S.A.C.N. Thomas, *Appl. Surf. Sci. Adv.*, 6 100184 (2021).
34. K.K. Rao, Y. Lai, L. Zhou, J.A. Haber, M. Bajdich, J.M. Gregoire, *Chem. Mater.*, 34, 899–910 (2022).
35. Q. Wang, J. Li, Y. Li, G. Shao, Z. Jia, B. Shen, *Nano Res.*, 1(5), 8751–8759 (2022).
36. Z.P. Wu, X.F. Lu, S.Q. Zang, X.W.D. Lou, *Adv. Funct. Mater.*, 30, 1910274 (2020).
37. M. Ďurovič, J. Hnát, K. Bouzek, *J. Power Sources*, 493, 229708 (2021).
38. D. Maity, G.G. Khan, Organic/inorganic metal halide perovskites for solar-driven water splitting: Properties, mechanism, & design, *Mater. Today Energy*, 37, 101407 (2023).
39. J. Xu, I. Amorim, Y. Li, J. Li, Z. Yu, B. Zhang, A. Araujo, N. Zhang, L. Liu, Stable overall water splitting in an asymmetric acid/alkaline electrolyzer comprising a bipolar membrane sandwiched by bifunctional cobalt-nickel phosphide nanowire electrodes, Energy, 2, 646–655 (2020).
40. S. Riyajuddin, K. Azmi, M. Pahuja, S. Kumar, T. Maruyama, C. Bera, K. Ghosh, Super-hydrophilic hierarchical Ni-Foam-Graphene-Carbon Nanotubes-Ni_2P-CuP_2 nano-architecture as efficient electrocatalyst for overall water splitting, *ACS Nano*, 15, 5586–5599 (2021).
41. A.L. Linsebigler, G. Lu, J.T. Yates, Photocatalysis on TiO2 surfaces: Principles, mechanisms, & selected results, *Chem. Rev.*, 95, 735–758 (1995).
42. T. Tachikawa, M. Fujitsuka, T. Majima, Mechanistic insight into the TiO2 photocatalytic reactions: Design of new photocatalysts, *J. Phys. Chem. C*, 111, 5259–5275 (2007).
43. V.M. Aroutiounian, V.M. Arakelyan, G.E. Shahnazaryan, Metal oxide photoelectrodes for hydrogen generation using solar radiation-driven water splitting, *Sol. Energy*, 78, 581–592 (2005).
44. G.K. Mor, O.K. Varghese, M. Paulose, K. Shankar, C.A. Grimes, A review on highly ordered TiO2 nanotube-arrays: Fabrication, material properties, & solar energy applications, *Sol. Energy Mater. Sol. Cells*, 90, 2011–2075 (2006).
45. M. Ni, M.K.H. Leung, D.Y. Leung, K. Sumathy, A review & recent developments in photocatalytic water splitting using TiO2 for hydrogen production, *Renew. Sustain. Energy Rev.*, 11, 401–425 (2007).
46. J. Nowotny, T. Bak, M.K. Nowotny, L. Sheppard, Titanium dioxide for solar-hydrogen. I. functional properties, *Int. J. Hydrogen Energy*, 32, 2609–2629 (2007).
47. E.L. Miller, D. Paluselli, B. Marsen, R.E. Rocheleau, development of reactively sputtered metal oxide films for hydrogen-producing hybrid multijunction photoelectrodes, *Sol. Energy Mater. Sol. Cells*, 88(2), 131–144 (2005).
48. K. Juodkazis, J. Juodkazytė, E. Jelmakas, P. Kalinauskas, I. Valsiūnas, P. Miečinskas, S. Juodkazis, Photoelectrolysis of water: Solar hydrogen–achievements & perspectives, *Opt. Express*, 18, A147–A160 (2010).
49. Jia, L.C. Seitz, J.D. Benck, Y. Huo, Y. Chen, J.W.D. Ng, T. Bilir, J.S. Harris, T.F. Jaramillo, Solar water splitting by photovoltaic-electrolysis with a solar-to-hydrogen efficiency over 30% Jieyang, *Nat. Commun.*, 7, 13237 (2016).
50. A. Fujishima, X. Zhang, D. Tryk, Heterogeneous photocatalysis: From water photolysis to applications in environmental cleanup, *Int. J. Hydrogen Energy*, 32, 2664–2672 (2007).
51. J.R. Bolton, S.J. Strickler, J.S. Connolly, Limiting & realizable efficiencies of solar photolysis of water, *Nature*, 316, 495–500 (1985).
52. A. Nada, M. Barakat, H. Hamed, N. Mohamed, T. Veziroglu, Studies on the photocatalytic hydrogen production using suspended modified photocatalysts. *Int. J. Hydrogen Energy*, 30, 687–91 (2005).

53. K. Maeda, D. Lu, K. Domen, Direct water splitting into hydrogen & oxygen under visible light by using modified TaON photocatalysts with d(0) electronic configuration, *Chem. A. Eur. J.*, 19, 4986–4991 (2013).

54. R. Molinari, C. Lavorato, P. Argurio, K. Szymański, D. Darowna, S. Mozia, Overview of photocatalytic membrane reactors in organic synthesis, energy storage & environmental applications, *Catalysts*, 9, 239 (2019).

55. Y. Jiang, C.Y. Toe, S.S. Mofarah, C. Cazorla, S.L.Y. Chang, Y. Yin, Q. Zhang, S. Lim, Y. Yao, R. Tian, Y. Wang, T. Zaman, H. Arandiyan, G.G. Andersson, J. Scott, P. Koshy, D. Wang, C.C. Sorrell, Direct hydrogen production from water/seawater by irradiation/vibration-activated using defective ferroelectric BaTiO3-x nanoparticles, *ACS Sustainable Chem. Eng.*, 11(8), 3370–3389 (2023).

56. A.A. Basheer, I. Ali, Water photo splitting for green hydrogen energy by green nanoparticles, *Int. J. Hydrogen Energy*, 44, 11564–11572 (2019).

57. G. Zhang, Z.A. Lan, L. Lin, S. Lin, X. Wang, Overall water splitting by Pt/g-C3N4 photocatalysts without using sacrificial agents, *Chem. Sci.*, 7, 3062–3066 (2016).

58. L. Lin, Z. Lin, J. Zhang, X. Cai, W. Lin, Z. Yu, X. Wang, Molecular-level insights on the reactive facet of carbon nitride single crystals photocatalysing overall water splitting, *Nat. Catal.*, 3, 649–655 (2020).

59. X. Tao, Y. Zhao, S. Wang, C. Li, R. Li, Recent advances & perspectives for solar-driven water splitting using particulate photocatalysts. *Chem. Soc. Rev.*, 51, 3561–3608 (2022).

60. M. Shi, R. Li, C. Li, Halide perovskites for light emission & artificial photosynthesis: Opportunities, challenges, & perspectives, *EcoMat*, 3, e12074 (2021).

61. K. Maeda, Z-Scheme Water splitting using two different semiconductor photocatalysts. *ACS Catal.*, 3, 1486–1503 (2013).

62. K. Sayama, K. Mukasa, R. Abe, Y. Abe, H. Arakawa, Stoichiometric water splitting into H_2 & O_2 using a mixture of two different photocatalysts & an IO_3^-/I^- Shuttle redox mediator under visible light irradiation, *Chem. Commun.*, 1, 2416–2417 (2001).

63. Q. Wang, T. Hisatomi, Q. Jia, H. Tokudome, M. Zhong, C. Wang, Z. Pan, T. Takata, M. Nakabayashi, N. Shibata, Y. Li, I.D Sharp, A. Kudo, T. Yamada, K. Domen, Scalable water splitting on particulate photocatalyst sheets with a solar-to-hydrogen energy conversion efficiency exceeding 1%, *Nat. Mater.*, 15, 611–615 (2016).

64. K. Takanabe, Photocatalytic water splitting: Quantitative approaches toward photocatalyst by design, *ACS Catal.*, 7(11), 8006–8022 (2017).

65. L. Han, M. Lin, S. Haussener, Reliable performance characterization of mediated photocatalytic water-splitting half reactions, *ChemSusChem*, 10, 2158–2166 (2017).

66. H. Kisch, D. Bahnemann, Best practice in photocatalysis: Comparing rates or apparent quantum yields? *J. Phys. Chem. Lett.*, 6, 1907–1910 (2015).

67. E. Parzinger, B. Miller, B. Blaschke, J.A. Garrido, J.W. Ager, A. Holleitner, U. Wurstbauer, Photocatalytic stability of single & few-layer MoS2, *ACS Nano*, 9, 11302–11309 (2015).

68. Z. Li, X. Meng, Z. Zhang, Recent development on MoS2-based photocatalysis: A review, *J. Photoch. Photobio. C*, 35, 39–55 (2018).

69. Y. Zhao, C. Ding, J. Zhu, W. Qin, X. Tao, F. Fan, R. Li, C. Li, A hydrogen farm strategy for scalable solar hydrogen production with particulate photocatalysts, *Angew. Chem. Int. Ed*, 59, 9653–9658 (2020).

70. A. Debierne, Recherches sur les gaz produits par les substances radioactives. Décomposition de l'eau, *Ann. Phys. (Paris)*, 2, 97–127 (1914).

71. R. Yamada, Y. Kumagai, R. Nagaishi, Effect of alumina on the enhancement of hydrogen production & the reduction of hydrogen peroxide in the γ-radiolysis of pure water & 0.4 M H2SO4 aqueous solution, *Int. J. Hydrogen Energy*, 36, 11646–11653 (2011).

72. F. Crumière, J. Vandenborre, R. Essehli, G. Blain, J. Barbet, M. Fattahi, LET effects on the hydrogen production induced by the radiolysis of pure water, *Radiat. Phys. Chem.*, 82, 74–79 (2013).

73. J.S. Southwortha, S.M. Pimblotta, R.M. Orrd, S.P.K. Koehlerb, A novel method for measuring the radiolysis yields of water adsorbed on ZrO2 nanoparticles, *Radiat. Phys. Chem.*, 174, 108924 (2020).

74. J. McGrady, S. Yamashita, S. Kano, H. Yang, A. Kimura, M. Taguchi, H. Abe, Radiolysis of water at the surface of ZrO2 nanoparticles, *Radiat. Phys. Chem.*, 209, 110970 (2023).

75. K.A.P. Kumar, G.A.S. Sundaram, S. Venkatesh, R. Gandhiraj, R. Thiruvengadathan, A Monte Carlo simulation study of L-band emission upon gamma radiolysis of water, *Radiat. Phys. Chem.*, 207, 110883 (2023).

76. I. Ali, G. Imanova, T. Agayev, A. Aliyev, O.M.L. Alharbi, A. Alsubaie, A.S.A. Almalki, A comparison of hydrogen production by water splitting on the surface of α-, β- & γ-Al2O3, *ChemistrySelect*, 7, e202202618 (2022).

77. I. Ali, H. Mahmudov, G. Imanova, T. Suleymanov, A.M. Hameed, A. Alharbi. Hydrogen production on nano Al2O3 surface by water splitting using gamma radiation, *J. Chem. Technol. Biotechnol.*, 98, 1186–1191 (2023).

78. I. Ali, G. Imanova, A.A. Garibov, T.N. Agayev, S.H. Jabarov, A.S.A. Almalki, A. Alsubaie, Gamma rays mediated water splitting on nano-ZrO2 surface: Kinetics of molecular hydrogen formation. *Radiat. Phys. Chem.*, 183, 109431 (2021).

79. I. Ali, T. Agayev, G. Imanova, H. Mahmudov, H. Musayeva, O.M.L. Alharbi, Siddiqui MN effective hydrogen generation using water-n-hexane-ZrO2 system: Effect of temperature & radiation irradiation time, *Mater. Lett.*, 340, 134188 (2023).

80. I. Ali, G. Imanova, T. Agayev, A. Aliyev, S. Jabarov, H.M. Albishri, W.H. Alshitari, A.M. Hameed, A. Alharbi, Seawater splitting for hydrogen generation using zirconium & its niobium alloy under gamma radiation, *Molecules*, 27, 6325 (2022).

81. I. Ali, G.T. Imanova, X.Y. Mbianda, O.M.L. Alharbi, Role of the radiations in water splitting for hydrogen generation, *Sustain. Energy Technol. Assess.*, 51, 101926 (2022).

82. J. McGrady, S. Yamashitaa, S. Kano, H. Yang, A. Kimura, H. Abe, H$_2$ generation at metal oxide particle surfaces under γ-radiation in water, *J. Nucl. Sci. Technol.*, 58, 604–609 (2021).

83. T. Kojima, K. Takayanagi, R. Taniguchi, S. Okuda, S. Seino, T.A. Yamamoto, Hydrogen gas generation from the water by gamma-ray radiolysis with pre-irradiated silica, *J. Nucl. Sci. Technol.*, 43, 1287–1288, 82 (2006).

84. S. Esnouf, A. Dannoux-Papin, E. Bossé, V. Roux-Serret, C. Chapuzet, F. Cochin, J. Blancher, Hydrogen generation from α radiolysis of organic materials in transuranic waste: Comparison between experimental data & storage calculations, *Nucl. Technol.*, 208, 347–356 (2022).

85. H.E. Sims, K.J. Webb, J. Brown, D. Morris, R.J. Taylor, Hydrogen yields from water on the surface of plutonium dioxide, *J. Nucl. Mater.*, 437, 359–364 (2013).

86. G.V. Buxton, Radiation chemistry, In Farhataziz, M.A.J. Rodgers (Edts.), *Principles & Applications*, Verlag Chemie Publishers (1987).

87. J.W.T. Spinks, R.J. Woods, *An Introduction to Radiation Chemistry*, 3rd ed., Wiley-Interscience publication (1990).

88. K.A. Domnanich, G.W. Severin, A model for radiolysis in a flowing-water target during high-intensity proton irradiation, *ACS Omega*, 7(29), 25860–25873 (2022).

89. I. Ali, G. Imanova, O.M.L. Alharbi, A.M. Hameed, M.N. Siddiqui, Recent updates in direct radiation water-splitting methods of hydrogen production, *J. Umm Al-Qura Univ. Appl. Sci.*, In Press (2023).

90. G.T. Imanova, Gamma rays mediated hydrogen generation by water decomposition on nano-ZrO2 surface, *MAMS*, 4, 508–514 (2021).

91. H. Sheng, L. Yu, Y. Jian-Hua, Y. Ying, in Nanotechnology for sustainable energy, *J. Am. Chem. Soc.*, 1140(9), 219–241 (2013).

92. M.R. Gholipour, C.T. Dinh, F. Bélandb, T.O. Do, Nanocomposite heterojunctions as sunlight-driven photocatalysts for hydrogen production from water splitting, *Nanoscale*, 7, 8187 (2015).
93. T. Agayev, G. Imanova, I. Ali, S. Aliyev, A. Aliyev, I. Faradj-Zade, G. Iskanderova, M. Tagiyev, S. Dzhafarova, A. Ahmadova, Radiation-heterogeneous processes in Zr and alloys Zr 1%Nb in contact with seawater, *Int. J. Adv. Eng. Manag.*, 4, 257–264 (2022).
94. G.T. Imanova, Kinetics of radiation-heterogeneous & catalytic processes of water in the presence of zirconia nanoparticles, *Adv. Phys. Res.*, 2, 94–101 (2020).
95. H. Kim, J.W. Seo, W. Chung, G.M. Narejo, S.W. Koo, J.S. Han, J. Yang, J.Y. Kim, S.-I. In, Thermal effect on photoelectrochemical water splitting toward highly solar to hydrogen efficiency, *ChemSusChem*, 2023, 16, e20220 (2017).
96. S. Yamaguchi, Y. Maegawa, K. Fujita, S. Inagaki, Hydrogen production from methanol-water mixture over immobilized iridium complex catalysts in vapor-phase flow re action, *ChemSusChem*, 14, 1074–1081 (2021).

5 Industrial production of hydrogen by water splitting

5.1 INTRODUCTION

The industrial creation of hydrogen by water splitting is very important because it is needed in real life [1]. At the moment, there are two main methods for industrial hydrogen production by water splitting, i.e., electrolysis and thermochemical procedures. Electrolysis is the most collective method for industrial hydrogen manufacture from water. It includes the use of electricity to split water into hydrogen and oxygen. There are two chief types of electrolysis, i.e., water electrolysis involving electrodes and solid oxide electrolysis involving operating at high temperatures (normally above 800°C) and utilizing steam for hydrogen manufacture. The thermochemical procedures comprise the usage of heat to drive chemical reactions that result in the decay of water to yield hydrogen. The selection of the method for industrial hydrogen generation depends on variables like energy source accessibility, efficacy, production scale, and purity necessities. Electrolysis is the most extensively utilized method, being commercially obtainable. The industrial generation of hydrogen is performed in a reactor, as discussed in this chapter. Besides, the challenges in industrial production of hydrogen and their solution are also discussed.

5.2 INDUSTRIAL REACTORS

The industrial generation of hydrogen by water splitting is a developing field, with constant efforts to advance efficacy, decrease costs, and encourage the use of sustainable energy resources to make the process more ecologically friendly. In both thermochemical and electrolysis procedures, the industrial reactors are planned for large-scale generation, and numerous variables are carefully considered. The most significant are materials, energy source, efficiency, safety, scalability, and purity control. The reactor materials should be corrosion-resistant and capable to withstand high temperatures, depending on the procedure. The selection of energy resources like solar power and electricity plays an important role in designing the reactors. The industrial reactors' goal is to optimize energy efficacy, maximizing hydrogen production while minimalizing energy intake. The reactor design should allow for stress-free scaling to meet variable demands for hydrogen. Industrial reactors should follow strict safety ethics, as the generation of hydrogen gas can be dangerous. The specified equipment is utilized to separate and purify hydrogen gas generated to meet the wanted purity standards for numerous applications.

DOI: 10.1201/9781003432364-6

There are various types of reactors used in the production of hydrogen at an industrial scale by water splitting. These reactors depend on the types of methods used. It is also discussed in this chapter that electrolysis and thermochemicals are the two main processes for hydrogen production from water. The most important reactors used at industrial levels in electrolysis are filter-press and membrane electrolyzers.

5.2.1 ELECTROLYSIS REACTORS

The filter-pressed electrolyzers with mono- or bipolar electrodes are typically utilized. The filter pressed electrolyzers with mono- or bipolar electrodes are regularly utilized for water electrolysis. The forced flow of the electrolyte between electrodes is normally utilized, or the design utilized in current industrial Zdansky-Lonza electrolyzers is used for gas filling decrease [2]. The electrolyzer's chief electrode is of corrugated thin metal through which remote electrodes of the metal grid are pressed. The electrolyzer contains numerous parts like the asbestos diaphragm, gas channel, corrugated electrode, electrode grids, gasket, and feed channel. The free gas bubbles formed on the grid make the main electrode side conceded away by electrolytes circulating via the channels made by the corrugations of the chief electrode. The grid cell size is chosen to be lesser than the asbestos fibers' regular length, which stops the demolition of the diaphragm. To augment the corrosion resistance of the cathode and reduce the overvoltage of hydrogen emission, the grid surface is stimulated by using a platinum or palladium layer.

All the measures at dipping energy intake (catalyst use, high temperatures, and electrode designs) are utilized in present electrolyzers. To achieve electrolytic hydrogen manufacture in the European Union, Thyssen Krupp electrolyzers and Simiens Silyzer electrolyzers are utilized. The electrolysis plants are furnished with membranes with a polytetrafluoroethylene matrix and an inorganic group $-SO_3H$, which have established the trade names Flemion, Nafion, Asiplex, etc. It is not essential to use aqueous solutions of acids, salts, or alkalis, but distilled water may be used as the initial material. This feature is utilized in Simien's electrolyzers. Another property of membrane electrolyzers is the utilization of dynamic catalysts for water degradation straight on the membrane surface (Figure 5.1). The platinum-based catalysts are used for this purpose. Consequently, it is probable to decrease the overvoltage of gas emission and ohmic voltage drops and decrease energy intake for gaining 1.0 nm^3 of hydrogen below 4.3 kW/h.

In the Silyzer electrolysis plant, water is provided by a pipe to the cathode and anode, and the discharge gases enter separators situated just above the electrolyzers. The gas enters the collectors and is collected for further applications. The electrolysis plants with power consumption of 10–20 MW are gathered from separate electrolyzers. These are categorized by an outcome of 2,000–4,000 m^3 of hydrogen/hour. Water feeding is less than 1.0 L/m^3 of hydrogen, with a gas purity of 99.95%. The electrolysis plant contains wash, dry, and post-treatment of hydrogen. The hydrogen amount in the gas is ~99.999% after cleansing. The working temperature is about 90°C with 1.8 V at a current density of around 1,000–2,000 A/m^2. The water-splitting voltage in these electrolyzers is 1.23 V.

Although conservative low-temperature electrolysis may be joined with all working reactors yet it will not be viable economically. The various potential procedures

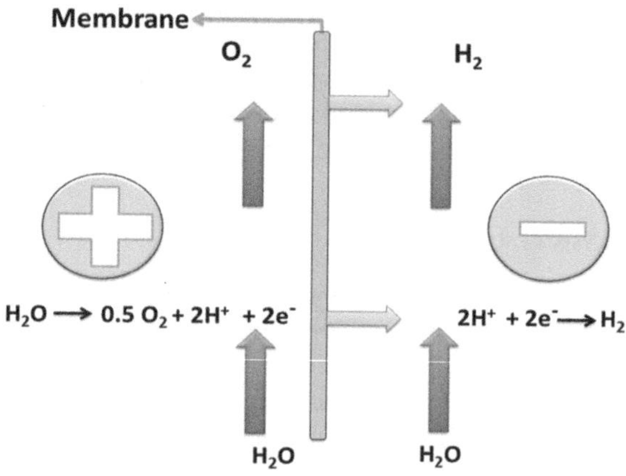

FIGURE 5.1 Schematic representation of a membrane electrolyzer.

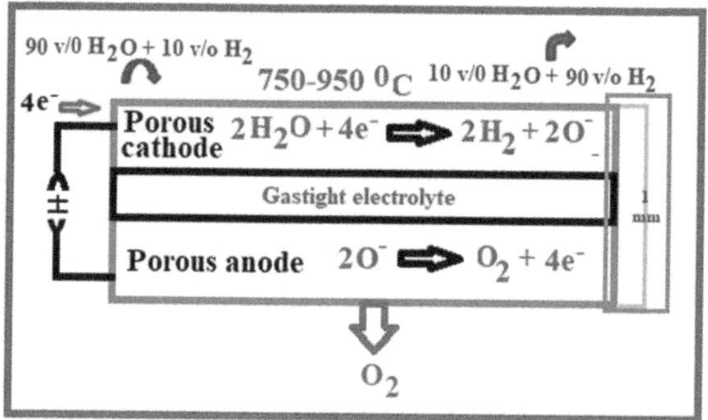

FIGURE 5.2 Schematic representation of a planar steam electrolysis cell.

for hydrogen manufacture like high-temperature steam electrolysis (HTSE) and other thermochemical procedures are recognized at higher temperatures. The utilization of high-temperature reactors for hydrogen generation grants a feasible choice as many of these procedures have higher efficacy than low-temperature electrolysis. The electricity feedback needed for electrolysis may be reduced by augmenting the temperature. This is due to the fact that the total energy needed for electrolysis in the vapor phase is decreased by the heat of vaporization that can be provided more inexpensively by thermal rather than electric source. Truly, at a high temperature of 800.0°C–1,000.0°C, electricity feedback might be around 35% lower than that of classical electrolysis. Additionally, the efficacy of electrical production at high temperatures is considerably good. The HTSE procedure is beneficial owing to its high complete thermal-to-hydrogen efficacy when joined with a high-efficacy power series. All the reactions proceed very fast at high temperatures. The steam-hydrogen mixture exiting from the load is passed via the separators to get hydrogen separated. The intake gas steam to HTSE cell comprises a 10% portion of hydrogen to preserve reduction in conditions and escape the nickel oxidation in the hydrogen electrode (Figure 5.2).

The HTSE cells can function at high current densities that permit huge hydrogen generation capacities. An electricity-to-hydrogen efficacy of around 90.0% could be attainable. Nevertheless, the life of hydrogen electrodes that are restricted by degradation needs further advancement.

5.2.2 Thermochemical Reactors

The thermochemical reactors for hydrogen generation from water are encouraging methods that utilize high-temperature chemical procedures to split water into hydrogen (H_2) and oxygen (O_2). These reactors are substitutes for the conventional methods of hydrogen generation. These are steam methane reforming and electrolysis. The thermochemical reactors provide the perspective for high efficacy and the use of different heat sources, comprising solar energy to carry out the reactions. The most significant thermochemical reactors are hybrid sulfur, cerium-based two-step cycle, sulfur-iodine (S-I) cycle, ferrite-based thermochemical cycle, etc. The key benefits of thermochemical reactors for hydrogen production comprise their potential for high-temperature heat input that can be supplied by numerous sources like concentrated solar energy and innovative nuclear reactors. This makes them the talented choice for sustainable and environmentally friendly hydrogen generation.

The S-I cycle was designed at General Atomics in the mid-1970s [3]. Iodine and sulfur dioxide are mixed with water in this process. These form sulfuric acid and hydrogen iodide in an exothermic reaction. These compounds are separated easily being immiscible. The sulfuric acid can be disintegrated at ~850°C giving oxygen and reprocessing the sulfur dioxide. The hydrogen iodide can be disintegrated at ~400°C, giving hydrogen and reprocessing iodine. The final reaction is the decay of water into hydrogen and oxygen. The complete procedure takes place in water alone. All the chemicals used are recycled with no effluents. Thermochemical water splitting is the change of water into hydrogen and oxygen by a sequence of thermally motivated chemical reactions. The straight thrombolysis of water needs a temperature above 2,500°C for noteworthy hydrogen generation. A thermochemical water-splitting cycle achieves a similar overall outcome utilizing a much lower temperature. The S-I cycle is the main example of a thermochemical cycle. It contains three chemical reactions, which figure out the breaking of water, as shown below:

$$I_2 + SO_2 + 2H_2O \rightarrow 2HI + H_2SO_4 \quad (120.0°C) \rightarrow (\text{Exothermic}) \qquad (5.1)$$

$$H_2SO_4 \rightarrow SO_2 + H_2O + \tfrac{1}{2}O_2 \quad (\sim 850.0°C) \rightarrow (\text{Exothermic}) \qquad (5.2)$$

$$2HI \rightarrow I_2 + H_2 \quad (\sim 400.0°C) \; (\text{Exothermic}) \qquad (5.3)$$

$$H_2O \rightarrow H_2 + \tfrac{1}{2}O_2 \qquad (5.4)$$

These reactions may be symbolized in a cycle (Figure 5.3).

One or more endothermic high-temperature chemical reactions provide the heat energy that enters a thermochemical cycle. One or more exothermic low-temperature processes are used to reject heat. In this process, all the chemicals used are recycled, except water. Most of the supplied heat is utilized to break sulfuric acid. The hydrogen iodide and sulfuric acid are formed in the exothermic reaction of water, sulfur dioxide, and iodine. The hydrogen is produced in the degradation of hydrogen.

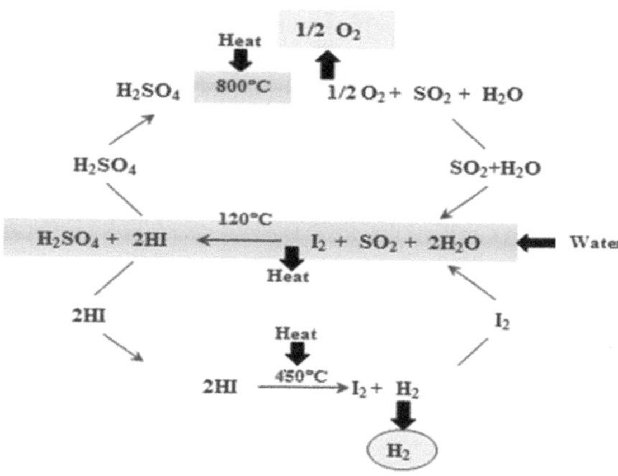

FIGURE 5.3   The chemical reactions in sulfur-iodine cycle reactor.

Driver et al. [4] developed a new reactor idea consisting of a counter-rotating ring receiver/reactor/recuperator. The researchers demonstrated the recovery of sensible heat in a cyclic process with this reactor. This reactor had two revolving beds of ferrite reactant material, rotating in reverse directions. The solar radiation imposes the front of the rotating rings. The intermediate rings were utilized to recuperate functional heat and the final part was utilized to complete the hydrolysis reaction. One of the benefits of this model comprised reactant support. Consequently, this was harmonious with concentration arrangements with flexible alignments relative to gravity. The reactor was planned to reach 32.2% thermal efficacy, 35% reaction extent, and 0.0412 mol/s hydrogen production rate with 15 kW solar power.

To make the reactors economical, solar energy is being used as solar-to-chemical energy change efficacy enhancement. It is expressed as a ratio between energy stocked in the product (Q_{ch}) and the incident solar rays on the absorber (Q_{solar}).

Many countries of the world have started to develop nuclear reactors for hydrogen manufacture at large scale. A typical nuclear hydrogen production [5] plant comprises a very high temperature reactor, a hydrogen production plant (HPP), an intermediate heat exchanger, and a power conversion system. The Institute of Nuclear Energy Technology, China, developed the high-temperature gas-cooled reactor (HTGR) in 2000. A number of reports on nuclear hydrogen production were released by the Idaho National Laboratory and the Idaho National Engineering Laboratory in 2006 and 2008. Three advanced reactor power cycle combinations were examined: (1) a supercritical CO_2-cooled reactor connected to a direct recompression cycle; (2) a helium-cooled reactor coupled to a direct Brayton power cycle; and (3) a fast reactor with sodium cooling coupled to a Rankine cycle. There is active research and development being done on nuclear hydrogen production. Several nations are at the forefront of this field, most notably the USA, Japan, Korea, and China. The feasibility of producing hydrogen on a big scale by nuclear means has previously been proven by Japan. These advancements would pave the path for the large-scale manufacturing of hydrogen from nuclear energy, which would quickly emerge as the sustainable worldwide fuel of the future.

The best reactor is that produces hydrogen at a large scale with low cost and no environmental pollution. Some of the important requirements for the best reactors are:

- **Temperature:** 750°C and 1,000°C are mandatory. Higher temperatures are favored.
- **Heat transfer:** Heat should be transported from a nuclear plant to a chemical set-up at a high temperature. The nuclear and the chemical plants should be protected from each other due to both having significant hazardous materials. This prerequisite levies predefined restraints on the reactor.
- **H_2 generation capability:** The main condition is to deliver heat at high temperatures.
- **HTGR:** Many variations of HTGR are available. These include a pebble-bed reactor and a hexagonal fuel-block reactor.
- **Advanced high-temperature reactor (AHTR):** This is a molten-salt-cooled modular reactor that utilizes a covered particle graphite matrix fuel. AHTR is parallel to an HTGR excluding high-pressure coolant that is substituted with a low-pressure molten salt. The permissible temperature may be slightly higher than for HTGR, and the coolant functions at atmospheric pressure.
- **Lead-cooled fast reactor:** This is a nitride-fueled and lead-cooled reactor. The working temperatures are slightly lower than those used in HTGR. Lead cooling is needed due to sodium boils at 883°C. This is considerably below the needed operating temperatures.
- **Graphene and nanomaterials as photocatalysts:** Graphene and other nanomaterials have attained notable attention as potential photocatalysts for water splitting. The application of nanomaterials as photocatalysts in water splitting has the prospective to discuss some challenges, comprising improved efficacy and reduced energy demand. These photocatalysts may be outstanding materials due to their extraordinary properties as already discussed in Chapter 4.

5.3 CHALLENGES IN THE INDUSTRIAL PRODUCTION OF HYDROGEN

Industrial hydrogen generation by water splitting faces numerous challenges, despite the promise of hydrogen as a clean energy carrier. These challenges are technical, economic, and environmental. The most significant challenges are cost, scale-up, energy input, efficiency, material compatibility, safety, purity, and environmental concern.

Water splitting needs a substantial amount of energy to break H_2O molecules into hydrogen (H_2) and oxygen (O_2). Most industrial processes trust electricity (electrolysis) or high-temperature heat (thermochemical methods). The sourcing of clean and sustainable energy for these processes, such as renewable electricity or high-temperature heat, can be exciting. The cost of producing hydrogen by water splitting is frequently higher. The capital and operational costs of water-splitting facilities, as well as the cost of electrolysis or high-temperature reactors, are required to be condensed for hydrogen to be economically competitive. The current water-splitting methods can have low energy change efficiencies compared to other hydrogen manufacturing methods. Improving the efficacy of electrolysis, and thermochemical reactors, is critical to reduce the energy input and cost. Water-splitting methods include the use of photocatalysts that can lead to material ruin and corrosion over time. Developing photocatalysts that can withstand these conditions is a main challenge. The transition from laboratory-scale procedures to large-scale industrial manufacture is challenging. Ensuring the scalability of water-splitting technologies while maintaining efficacy and cost-effectiveness is critical. Many applications need high-purity hydrogen. Ensuring hydrogen purity produced via water-splitting processes can be technically challenging. Hydrogen is a highly flammable gas and has specific safety deliberations for production, storage, and transportation. Confirming safety while working with hydrogen on an industrial scale is a challenge. The environmental footprint of hydrogen manufacture should be minimized. For example, if the electricity utilized for electrolysis is generated from fossil fuels, it should not be a green option. The use of renewable energy sources for electricity or heat generation is a critical aspect [6].

The unfavorable thermodynamics of the water-splitting reaction is a major challenge for photocatalytic water splitting. The inactive kinetics of the water-splitting reaction is another challenge for photocatalytic water splitting. The occurrence of dissolved oxygen can hinder the hydrogen manufacture reaction and can be a safety issue when mixed with hydrogen in definite concentrations in photo-biological hydrogen manufacture. The backward reaction can decrease the efficacy of water splitting reaction. The side reaction can decrease the efficacy of the water-splitting reaction. Low hydrogen production rates are a challenge for photo-biological hydrogen production. Water splitting also produces oxygen, which can rapidly hinder hydrogen manufacture reaction and can be a safety issue when mixed with hydrogen in some concentrations in photo-biological hydrogen production [7]. Besides the technical challenges, the distribution of water-splitting techniques is influenced by regulations and policies. The supportive policies and incentives, like subsidies or carbon pricing mechanisms, can augment the adoption of clean hydrogen techniques. Establishing a hydrogen infrastructure, including production, distribution, and refueling stations, is a noteworthy challenge, especially in regions where hydrogen is not yet widely adopted.

5.4 SOLUTION TO THESE CHALLENGES

Addressing the challenges in the industrial production of hydrogen by water splitting requires a mixture of technological advancements, policy support, and research efforts. Some potential solutions to the challenges are clean energy sources, cost reduction, efficiency improvements, scalability, environmental impact mitigation, photocatalysts development, safety protocols, infrastructure development, purity standards, research and collaboration, and regulatory and policy support. The transition to clean and sustainable energy sources for water splitting like renewable electricity (solar, wind) and nuclear power is needed. Using low-carbon or carbon-free energy sources reduces the carbon footprint of hydrogen production. These should be considered during hydrogen production. Advances in catalysts, materials, and reactor designs should advance to improve the overall efficacy of hydrogen manufacture. The development of photocatalysts that can withstand severe conditions in water-splitting processes, such as high-temperature and chemically aggressive environments should be developed. The economies of production should be measured to reduce the capital and operational costs of hydrogen manufacture. The focus should be given to the scaling up of water-splitting techniques for industrial usage. The pilot projects and associations between research institutions and industry can help bridge the gap between lab-scale experiments and commercial manufacture.

The environmental impact of hydrogen manufacture should be minimized. The utilization of life-cycle analysis to measure the carbon footprint of diverse production methods is important. The prioritization of the use of low-carbon energy sources should be stressed. The development and execution of rigorous safety protocols for handling, storing, and transporting hydrogen should be highlighted. The advances in safety technology, leak detection, and emergency response systems can improve the safety of hydrogen handling. These processes should be considered during hydrogen manufacture. The establishment and implementation of standards for hydrogen purity should be fixed. The development of effective purification methods can aid in meeting purity necessities. A good investment in the development of hydrogen infrastructure, including hydrogen refueling stations, pipelines, and distribution networks, is beneficial to solve many problems. The public-private partnership can advance the infrastructure. The collaboration between industry, research institutions, and government agencies should be encouraged to advance water-splitting technology. The continued research and development efforts are crucial to drive improvement and address the lying challenges. Government incentives and subsidies can also make clean hydrogen more economically inexpensive. Governments can play a vital role by implementing supportive policies and regulations that incentivize clean hydrogen generation and utilization. This includes carbon pricing mechanisms, tax incentives, and emissions reduction targets. Briefly, these challenges can be solved by a concentrated effort from multiple stakeholders, including governments, research institutions, industry leaders, and environmental organizations. The collaboration and long-term commitments to clean hydrogen production are critical for realizing hydrogen's potential as a clean and sustainable energy carrier.

REFERENCES

1. I. Ali, G. Imanova, O.M.L. Alharbi, A.M. Hameed, M.N. Siddiqui, Recent updates in direct radiation water-splitting methods of hydrogen production, *J. Umm Al-Qura Univ. Appl. Sci.*, In Press (2023).
2. T. Smolinka, H. Bergmann, J. Garche, M. Kusnezoff, The history of water electrolysis from its beginnings to the present Electrochemical Power Sources, In T. Smolinka, J. Garche (Edts.), *Fundamentals, Systems, and Applications: Hydrogen Production by Water Electrolysis*, Elsevier, 83–164 (2021).
3. G.E. Besenbruch, General atomic sulfur-iodine thermochemical water-splitting process, *Am. Chem. Soc., Div. Pet. Chem.*, 271, 48 (1982).
4. R.B. Diver, J.E. Miller, M.D. Alledorf, N.P. Siegel, R.E. Hogan, Solar thermochemical water-splitting ferrite-cycle heat engines, *J. Sol. Energy Eng.*, 130, 041001 (2008).

5. R. Elder, A. Ray, Nuclear heat for hydrogen production: Coupling a very high/high temperature reactor to a hydrogen production plant, *Prog. Nucl. Energy*, 51, 500–525. (2009).
6. S.K. Dash, S. Chakraborty, D. Elangovan, A brief review of hydrogen production methods and their challenges, *Energies*, 16, 1141 (2023).
7. C. Bie, L. Wang, J. Yu, Challenges for photocatalytic overall water splitting, *Chemistry*, 8, 1567–1574 (2022).

Part II

Hydrogen storage and transport

6 Hydrogen storage

6.1 FUNDAMENTALS INTRODUCTION

Hydrogen storage is a serious issue because it requires energy to store and discharge hydrogen for further use. However, the investigators tried to find the rooms for hydrogen storage by using a variety of approaches. Hydrogen can be stored as a gas or a liquid and utilized for some beneficial purposes. High-pressure containers (5,000.0–10,000.0 psi container pressure) are often needed for gas storage. Since hydrogen has a boiling point of −252.8°C at one atmosphere of pressure, storing it as a liquid needs cryogenic temperatures. The need to use hydrogen for on-board energy storage in zero-emission vehicles is driving the creation of novel storage technologies that are better suited for this new purpose. The main obstacle is extremely low boiling point of hydrogen, which is 20.268 K (−252.882°C or 423.188°F). These low temperatures demand a substantial amount of energy. A big problem is storing adequate hydrogen within a car to reach a driving range of more than 300 miles. In terms of energy per unit of weight, hydrogen has about three times as much as gasoline (120 MJ/kg vs. 44 MJ/kg). In contrast, the situation is flipped when measured in terms of volume (8 MJ/L for liquid hydrogen (LH_2) against 32 MJ/L for gasoline). To accommodate the entire platform of light-duty cars, on-board hydrogen storage in the 5–13 kg H_2 range is needed. However, the choice of the best hydrogen storage method depends on various factors such as storage capacity, energy efficiency, safety, cost, and application requirements. Currently, there is no single storage method that is universally considered the best. Instead, different storage methods are used based on specific applications and conditions. It might be possible to store bigger amounts of hydrogen in smaller volumes at low pressures and temperatures near to room temperature by storing it in solids. The hydrogen molecule is dissociated into hydrogen atoms in a metal hydride lattice, allowing for volume-based store densities higher than LH_2.

Generally, the efficacy of storage is determined in two ways: the gravimetric density (GD), the percentage of hydrogen weight stored of whole weight of the structure (hydrogen with container), and the volumetric density (VD) which is the stored mass of hydrogen per unit volume of the system. Both variables are important because a hydrogen storage device should be compact and light for practical applications. Therefore, the probable storage systems may be assessed by plotting GD and VD in a Cartesian diagram. Even the existing US Department of Energy's (DOE) goals are stated in terms of VD and GD (green crosses in Figure 6.1) [1].

The ideal match between the dots of different sizes for physical sorption in nanotubes is represented by the orange line in the picture. The assessment of LH_2 for large nanotubes is where the line sways. Typically, the region below this line contains nanostructured physical sorption-based graphitic formations. The optimum physical sorption in graphene multilayers with space roughly doubling that of graphite (and density roughly halving) is indicated by the obliquely shaded strip. In this case,

DOI: 10.1201/9781003432364-8

various storage densities were matched to various pressures and temperatures. The parallel dark red stripes represent graphene that has been decorated or functionalized. These systems have typically only been taken into account at the layer-level. Only the GD is accurately established for this reason, but the VD range has been roughly determined while taking into account the variable inter-layer space that is 2–4 times that of graphite. The VD for chemical sorption in multilayers has been calculated using the same standard (blue rectangle). For this system, the GD has a sharp right edge that corresponds to the maximal packing with 1:1 stoichiometry of C and H (8%). As shaded areas in red, violet, and green, the storage properties of the systems are based on substances other than graphene (dissimilar metal hydrides, including MgH2), hydrocarbons, N- and B- hydrides). Green stars denote the DOE targets (for 2015 and the long term). The lines with constant density are gray. There are many methods of hydrogen storage but this chapter describes only those methods which may be used in real practice in the future. Out of these some important hydrogen storage methods are described in the following subsections.

6.2 COMPRESSED GAS

Hydrogen gas can be put in storage under great pressure in particularly designed containers. This way requires sturdy and frivolous containers. Condensed gas storing is a comparatively established method and has been utilized for hydrogen storing in numerous applications. Nevertheless, it needs high-pressure vessels, which can be large and have restricted storage capability. The safety considerations are also important due to the high-pressure conditions. The condensed hydrogen in containers at 350.0 bar (5,000.0 psi) and 700.0 bar (10,000.0 psi) is utilized in vehicles, based on type IV carbon-composite technique [2]. Car producers have been making this solution, such as Honda or Nissan.

The hydrogen storage tanks are of four types. (i) These tanks are made of metal and used to store hydrogen in industries. Normally, hydrogen is stored at 200–300 bar pressure. (ii) These tanks are, of course, of metals but are of liner hoop-draped joined cylinders. These types of tanks are used in industries for storage purposes. (iii) These tanks are completely draped amalgamated ones with a metal liner (aluminum) as a permeation barrier for hydrogen. [3]. These types of tanks are better in comparison to the first and second types in terms of 25%–75% mass gain with pressure till 450 bar. Consequently, these types of tanks are better for vehicle purposes [4]. However, challenges associated with these tanks are pressure cycling tests at 700 bar [3]. (iv) These tanks are completely wrapped combined tanks with a high-density plastic lining. These types of tanks are used when hydrogen is stored at high pressure, i.e., up to 1,000 bar [5]. It was observed that the first and second types are not suitable for transportation due to their heavy weights while the third and fourth types of tanks are costly due to the use of costly materials such as carbon fibers, plastics, etc.

New types of tanks may be termed type V, which were developed by Composites Technology Development Inc., USA, in 2010. Although it is meaningfully lighter than type IV containers, type V vessels do not work at high pressures [6]. Of course, these types of tanks were lighter than the fourth type but were not quite good in terms of hydrogen storage at high pressure. Nowadays, the vehicle industries need hydrogen to be stored at about 350–700 bar pressure and, hence, types III and IV are being used. The most common name industries are the Hyundai Tucson, Toyota Mirai, and Honda Clarity. Type IV tanks can provide about 5.2–5.5 wt% hydrogen, which corresponds to 18.0 and 28.0 g/L volumetric hydrogen capacities, correspondingly [7]. It may be concluded that types I and II are useful for industrial purposes while types III, IV, and V may be used for motor vehicles and aircraft in the future. However, there is still a need to improve the design, weight, and cost of these tanks.

6.3   LIQUID HYDROGEN (LH$_2$)

Hydrogen can be liquefied by cooling at a very low temperature (−253°C) called cryogenic storage. LH$_2$ has a high energy density in comparison to compressed gas, permitting more hydrogen to be stored in a specified volume. Nevertheless, it needs cryogenic settings and particular storing containers, which can be costly and present challenges in terms of boil-off and isolation. The LH$_2$ containers for cars, produced by BMW Hydrogen 7, Japan, have an LH$_2$ store site in Kobe port. The BMW company manufactured dual-fuel inner combustion engine vehicles utilizing LH$_2$. However, they have certain limitations such as high cost and high pressure of LH$_2$ in running engine due to heat generation of the engine. Such types of tanks may be useful in aircraft and other space purposes. Hydrogen is converted into liquid by dipping its temperature to −253.0°C, analogous to liquid natural gas that is kept at −162.0°C. A possible efficacy damage of only 12.79% can be attained, or 4.26 kW/h/kg out of 33.30 kW/h/kg [8]. The tanks for liquid H$_2$ are designed keeping in mind the reduced surface area to control the surrounding heat flow to the liquid H$_2$. The disadvantage of this mode of storage is the need for high energy for the liquefaction of gaseous hydrogen. The energy needed for liquefaction is about 25%–40% of hydrogen energy content. This is due to the fact that the theoretical volume densities of H$_2$ at its boiling point and atmospheric pressure is 70.0 g/L while it is in the range of 24–40 g/L in the compressed situation [9]. Furthermore, the hydrogen storage tanks need to be insulated, which increases the cost of the tanks. Sometimes, double-walled tanks with vacuumed facility are made, which increase the cost of the tanks.

The cryo-compression shows exciting cost benefits and storing costs (price per vehicle), which are essentially the lowermost when likened to any other methods [10]. For instance, a cryo-compacted hydrogen structure would cost $0.15 per mile (with the cost of fuel and all other related costs), while normal gasoline automobiles price between $0.08 and $0.10 per mile. Similar to liquid storing, cryo-compacted utilizes cold hydrogen (20.30 K and slightly above) to achieve a high energy density. Nevertheless, the foremost variance is at hydrogen heat-up owing to heat transmission with the atmosphere ("boil off"); tank is permitted to attain up to 350 bar pressure, a much higher pressure range. Consequently, it uses more time prior to hydrogen exhausts. In most situations, sufficient hydrogen is utilized by the car to retain the pressure well below the expelling limit. Thus, it has been established that a high driving range could be attained with a cryo-compressed tank, i.e., more than 650 miles (1,050 km) were driven with a full tank riding on a hydrogen-fueled engine of Toyota Prius.

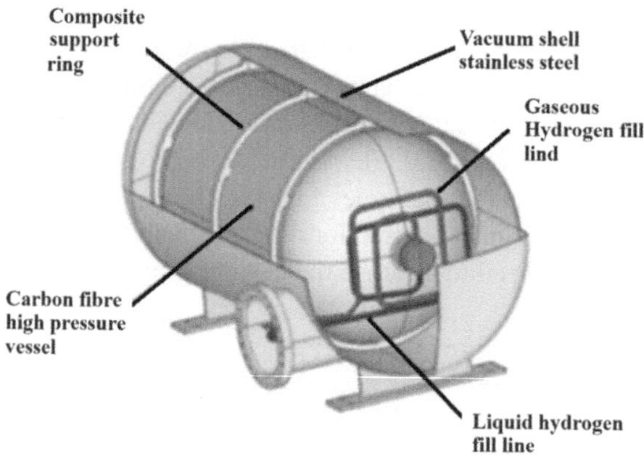

FIGURE 6.1 A schematic representation of a compressed cryogenic storage tank [reproduced with permission from ref. 13].

The insulated tanks are used to collect cryogenic hydrogen at high pressures. This improves the volume hydrogen storage capability and safety in comparison to the compressed cryogenic hydrogen. The volumetric hydrogen store ability is augmented from 70 to 240 g/L at 1–240 bar [7]. The insulated tanks allow high pressure and cover the dormancy time. This resulted in an augmented store density. These tanks may be a little inexpensive due to the low pressure (<300 bar) requirement in compression to the tanks (700 bar) used in LH_2 storage. These types of tanks may meet the needs of the US DOE system goals for automobile uses, i.e., hydrogen loss in dormancy of low daily driving and volumetric and gravimetric hydrogen capabilities [11]. It has been reported that these tanks are better than type three after a computation study in buses [12]. In these types of tanks, no loss of hydrogen was obtained even after a week dormancy period. As of 2010, the BMW Group has begun a detailed components and system-level authentication of cryo-compacted vehicle store on its means to a marketable product [10]. Aceves et al. [13] developed cryogenic pressure tanks with 2–3 times additional fuel capacity than normal ones (Figure 6.1). These tanks were tested in vehicles and found satisfactory. In 2012, BMW showed an initial cryo-compressed hydrogen storage method but the cost was quite high. As per the press resale by BMW on February 27, 2023, BMW iX5 Hydrogen is in progress in BMW Group's pilot plant at its Research and Innovation Center (FIZ), Munich, Germany. The research is in progress to study and establish the full possibility of the method.

6.4 UNDERGROUND HYDROGEN STORAGE

Hydrogen can be put in storage underground in salt caverns or exhausted oil and gas reservoirs. This method includes hydrogen compression and sending into the storing formations. Underground storage can provide extensive storage capability, but it needs appropriate underground construction and set-up. The research is going on in this direction and expansion is constantly discovering new storage approaches and resources to overcome the limits of current methods. The optimum hydrogen storage way may vary depending on the specific use, i.e., for stationary energy storage, industrial processes, transportation, etc. A unique approach is a blend of diverse storage methods to meet the requirements.

As discussed above, the underground hydrogen is stored in salt domes caverns, and exhausted oil and gas reservoirs [14]. The large amounts of hydrogen have been deposited in caves for several years without any problems. The storing of great amounts of LH_2 in the ground can work as a grid energy store. The round-trip efficacy is about 40% vs. 75%–80% for pumped hydro energy storage (PHES). European staff developed large-scale hydrogen storage using a salt cavern store, an electrolyzer, and a combined-cycle power plant.

According to the 2013 European project Hyunder, extra 85 caverns are needed to store wind and solar energy because pumped hydroelectric energy storage (PHES) and compressed air energy stirage (CAES) cannot handle it. According to a study done in salt caverns in Germany, the country's excess power (7% of all renewable energy produced by 2025 and 20% by 2050) will be converted to hydrogen and stored underground. By 2025, around 15,500,000 m^3 gas caverns would be required and, by 2050, approximately 60 caverns, or almost one-third of the total number of gas caverns in Germany, will be required. According to studies done by Sandia Labs in the USA, depleted oil and gas fields might be used to store vast amounts of hydrogen that is produced through renewable sources.

Chen et al. [15] determined the techno-economic chances of hydrogen geologic storage in salt caverns, depleted gas reservoirs, and saline aquifers in the Intermountain-West (I-WEST) area. The authors reported that the geologic hydrogen store may provide at least 72% energy intake of I-WEST area in 2020. The per kg of hydrogen storage costs reported were $3.27, $2.50, and $1.15 in saline aquifers, salt caverns, and depleted gas reservoirs, respectively. This type of study may be used to explore the hydrogen storage cost in other parts of the world. Lin et al. [16] described a net present value (NPV) assessment outline for geological hydrogen storing that integrated the updated techno-economic examination and market-based processes. As per the authors, the developed NPV outline may be utilized for project risk management by determining the key cost factors for store possessions. Ghaedi et al. [17] attempted to formulate a maximum height of hydrogen column that evades capillary breakthrough into caprock above saline aquifers and exhausted gas reservoirs. As per the results reported a higher dip angle of formation may reduce the stored mass of hydrogen. The other aspects of geological hydrogen storage have been discussed by various workers such as legal conditions [18], feasibility and storage capacity [19–32], geological impact, and implications [23–26]. Hassanpouryouzband et al. [27] wrote a review article on the storage of hydrogen in geological storage. Muhammed et al. [28] also presented a review article on hydrogen underground storage with emphasis on geological site insight, affecting factors, and future perspectives. Amirthan and Perera [29] presented a review article on hydrogen storage underground in Australia. The authors attempted to address future directions to identify significant gaps in using geological formations for the storage of hydrogen. Interested readers should consult these articles.

6.5   STORAGE BY ADSORPTION

Adsorption is one of the important phenomena for many applications. Hydrogen has been stored in various substances and the most important are metal hydrides, activated carbon (AC), zeolites, polymers, and other nanoparticles. The hydrogen storage on these are discussed in the following sub-sections.

6.5.1   METAL HYDRIDES

The metal hydrides can sorb and discharge hydrogen via chemical reactions and, consequently, are considered good approach for hydrogen storage. Metal hydrides offer the advantage of high hydrogen storage capacity and relatively low operating pressures. However, some metal hydrides require high temperatures for hydrogen release, which can affect the overall efficiency and practicality of the system. Metal hydrides are compounds that consist of a metal atom or ion bonded to one or more hydrogen atoms. They can store and release hydrogen gas through a process called hydrogen absorption and desorption. Hydrogen generation using metal hydrides typically involves two main steps: hydrogen absorption and hydrogen desorption. Metal hydrides can absorb hydrogen gas by forming chemical bonds with the hydrogen atoms. The absorption process is typically exothermic, meaning it releases heat. The metal hydride acts as a sponge, soaking up the hydrogen gas. The absorption capacity of a metal hydride depends on factors such as temperature, pressure, and the composition of the metal hydride.

When the metal hydride needs to release the stored hydrogen, it undergoes a desorption process. This process is typically endothermic, requiring an input of heat to break the chemical bonds between the metal and hydrogen atoms. The released hydrogen gas can then be collected and used for various applications, such as fuel cells or as a feedstock for industrial procedures. Metal hydrides have been extensively studied for hydrogen storage and generation due to their high hydrogen storage capacity, stability, and relatively low operating pressures compared to compressed hydrogen gas. Different metal hydrides exhibit varying properties, such as absorption and desorption kinetics, thermodynamics, and stability. Some common examples of metal hydrides used for hydrogen generation are as below.

- **Sodium alanate (NaAlH4):** Sodium alanate is a well-known metal hydride with a high hydrogen storage capacity. It can release hydrogen at moderate temperatures (~100°C–150°C).
- **Magnesium hydride (MgH2):** Magnesium hydride is another widely studied metal hydride for hydrogen storage. It has a high gravimetric and volumetric hydrogen density. However, it typically requires higher temperatures (~300°C–400°C) for hydrogen release.
- **Ammonia borane (NH3BH3):** Ammonia borane is a solid compound that can release hydrogen gas upon heating. It has attracted attention as a potential hydrogen storage material due to its high hydrogen content and relative stability.

- **Lithium aluminum hydride (LiAlH4):** Lithium aluminum hydride is a powerful reducing agent and can release hydrogen gas when treated with suitable catalysts.

Other hydrides for hydrogen store include complex metal hydrides, which often comprise sodium, lithium, calcium, aluminum, or boron, as well as normal hydrides of magnesium [30] or transition metals. Low activity (big safety) and high hydrogen store densities are provided by the hydrides used for storage applications. Lithium hydride, sodium borohydride, lithium aluminum hydride, and ammonia borane are the leading contenders. The first industrialized products, based on magnesium hydride, are being developed by a French company called McPhy Energy and have already been sold to several significant customers including Iwatani and ENEL. The amount of H_2 they transport and the reversibility of the storage method are once again the ongoing issues. At room temperature and pressure, some are liquids that are simple to fuel, while others are solids that can be processed into pills. Although these materials have a normal energy density, their specific energy is frequently lower than that of the most popular hydrocarbon fuels. Doping with activators is an alternative strategy for reducing dissociation temperatures. This technique has been utilized to create aluminum hydride, however it is unappealing due to the difficult synthesis [31].

Researchers continue to explore and develop new metal hydride systems with improved properties for efficient and safe hydrogen storage and release. Wang et al. [32] developed a metal hydride tank for hydrogen storage with diagonal baffles showing the small hydrogen refueling period due to the highest cooling water velocity. The practicality and financial viability of the hydrogen production plants were assessed by Kotowicz et al. [33], who presented the findings of their economic examination along with sensitivity analyses for different economic aspects. The case study under analysis demonstrates that alkaline electrolyzers have favorable economic characteristics across a wide range of parameters, but a polymer electrolyte membrane (PEM)-based plant is more resistant to changes in the electricity price, which is the major cost factor in hydrogen production. An experimental research of low-pressure store techniques using porous metal hydride containers enriches the study. In the same volume and pressure, the tanks' efficiency in comparison to pressurized hydrogen tanks is 10.2. The employment of them in a distributed hydrogen generation idea is suggested as a resolution. The effectiveness of the tanks is studied in relation to a method for calculating the avoided energy usage. Incorporating metal hydride tanks results in avoided energy usage of 1.33–1.37 kWh/kg of hydrogen, depending on number compressor stages. The techniques outlined in this study are applicable to all green hydrogen facilities and are universal. Giorgio et al. [34] described a hybrid energy storage system made of lithium-ion battery pack and metal hydride. The authors reported that the method was capable to control passively the temperature of the pack while augmenting storage energy density at the same time. This model gave 16% of total hydrogen mass by the fuel cell stack during working, which may be a significant augment of hydrogen storage for vehicles.

Chang et al. [35] proposed a thermochemical material to control of release and storage reaction heat, with a new sandwich bed for good hydrogen and thermal storage performance. Jana et al. [36] developed a metal hydride tube bundle reactor for hydrogen storage. The authors described it as effective with store capacity (~385 g hydrogen). Ye et al. [37] described a simple straight-tube heat exchanger surrounded by phase change material to increase hydrogen store enactment. The authors reported the tank as an alternative to control the heat transfer. Zhan et al. [38] described an approach for quick hydrogen charge and discharge of high-density hydrogen stores by the experimental and simulation validation of metal hydrides. The authors reported optimization of structure and mass and energy transfer concerns for innovative energy store purposes. Wang et al. [39] developed a design of a disc mini-channel reactor based on the metal hydride with good hydrogen storage efficacy. The authors used 3D Computer and Solution (COMSOL) models to optimize the reaction performance. A good performance was achieved at 0.9 mm height of heat transfer layer with 5 mm metal hydride layer height, relating to maximum hydrogen storage quantity per unit height 0.17 g/mm.

Disli et al. [40] also carried out a numerical-based study on metal hydride hydrogen storage tanks. The modeling was carried out in three scenarios, i.e., external convection, passive thermal management, and hybrid approach, including phase change material layer and an internal cooling pipe. Among all scenario three provided a metal hydride bed area at a lower temperature in comparison to scenario two. Elkhatib and Louahlia [41] carried out modeling studies of hydrogen tanks by investigating the thermal behavior of several cooling/heating modes. The authors reported a high sorbed hydrogen flow rate by decreasing temperature. The charge was amplified by 1.6% by reducing the refrigeration temperature to 1°C. Similarly, Afzal et al. [42] also carried out a simulation study on hydrogen storage reactor based on metal hydride. The authors established thermal resistance network analysis for limited heat transfer by conduction. Li et al. [43] presented a mathematical model discussing hydrogen sorption on $LaNi_{4.7}Co_{0.3}$ metal hydride as a good material. The model was the Jonhson-Mehl-Avrami-Kolmogorov model and the kinetic constant of the hydriding phenomenon was established. Aruna et al. [44] discussed modeling, PID controller design, and system documentation for the dynamics discharge of hydrogen storage on a metal hydride bed. The authors proposed a Fuzzy-PID controller for better control, which produces improved time domain response and better performance, as claimed by the authors. There are some reviews on hydrogen storage on metal hydrides. Interested readers can consult them [45–52].

6.5.2 ACTIVATED CARBON (AC)

Amorphous carbon compounds with a large apparent surface area make up ACs, which are extremely porous. By augmenting the apparent surface area and optimizing the pore width to around 0.7 nm, it is possible to boost the hydrogen physisorption in these materials [53]. Due to their ability to be produced from waste materials, like cigarette butts, which have shown noteworthy potential as precursor materials for high-capacity hydrogen store materials. These materials are of specific attention [54,55]. The large surface area and porous structure of AC allow it to absorb hydrogen molecules to its surface. Because of the reversibility of this adsorption process, AC has the potential as a material for storing and releasing hydrogen gas. To adsorb and retain hydrogen molecules, the surface area and pore size distribution are essential. Through weak Van der Waals interactions, hydrogen molecules can cling to AC surfaces. The AC has multiple holes and crevices, which give hydrogen molecules lots of places to bind with one another. AC's ability to store hydrogen is influenced by several variables, including its specific surface area, pore size distribution, temperature, and pressure. Thermodynamic and kinetic variables control the adsorption of hydrogen on AC. Adsorption capacity is often increased by higher pressures and lower temperatures, although pore structure and surface features may have an impact on the rate of adsorption and desorption. Reduced pressure and raised temperature can be used to release the hydrogen that has been held. The desorption of hydrogen from the surface of AC is extremely simple due to the weak Van der Waals interactions. The liberated hydrogen can then be utilized in a variety of systems, including fuel cells. Although AC has shown promise for storing hydrogen, there are still issues to be solved. These include boosting the kinetics of adsorption and desorption, improving the material's characteristics for increased storage capacity, and guaranteeing efficient and safe hydrogen release.

Pedicini et al. [56] prepared AC from *Posidonia oceanica* and Wood chips for hydrogen storage. The authors reported hydrogen sorption under cryogenic conditions (from 0.1 wt% to over 5 wt%). Doğan et al. [57] used tangerine peel to make AC with increased surface area by treating $ZnCl_2$ and KOH chemicals. The ACs prepared by $ZnCl_2$ had more hydrogen storage capability in comparison to the AC made by KOH. Pirsaheb et al. [58] reported green synthesis of AC nanocomposites with CuO, SnO_2, and Fe_2O_3. The AC with SnO_2 showed good hydrogen sorption capacity. Li et al. [59] used waste peanuts to obtain ultra-high specific surface area AC for hydrogen sorption. The authors reported a quite good capacity of AC for hydrogen storage. Çetingürbüz and Turkyilmaz [60] produced AC from lithium activation using pyrolyzed tree species. The authors used the developed material for and used for hydrogen sorption. Musyoka et al. [61] prepared AC coal fly ash by acid treatment for hydrogen storage. The obtained AC was with meso-/macropores and 46.19–81.20 m^2/g as surface area. The hydrogen stored was found to be 1.35 wt% at 77 k and 1 bar pressure.

Besides, the normal AC, many researchers prepared composite and nano-AC and evaluated it for hydrogen storage. Kostoglou et al. [62] prepared nanoporous polymer-based AC for hydrogen storage (adsorption). The polymer used was nano-porous polymer-/polyanilin. The authors reported (~2,200 m^2/g) surface area and (~1.0 cm^3/g volume of AC. The hydrogen sorption was in cryogenic conditions, i.e., a hydrogen uptake of ~5.5 wt% at 77 K and ~60 bar with a ~8.3 kJ/mol heat of sorption at zero reporting. Luo et al. [63] prepared composite AC in situ by one-step synthesis. The material synthesized was 1.64 cm^3/g surface area and with hydrogen sorption to 6.93 wt% at 40 bar and 77 K. Wu et al. [64] developed AC with metal-organic frameworks (MOFs). The authors reported more hydrogen storage than the normal AC. Morandé et al. [65] modified AC with doping boron and nitrogen to augment the hydrogen sorption. The authors reported the highest hydrogen capacity of this material *ca.* 2.34% at 0.93 bar. Chibani et al. [66] studied the effect of material, phase change, and inconstant mass flow rate of hydrogen sorption on AC. Hydrogen was controlled by the mass flow rate and it was proportional to the rise in the mass flow rate. Li et al. [67] modified nano-tubes of cup-shaped and used for hydrogen sorption. The authors described that nano-tubes of cups treated with potassium hydroxide may be the potential material for good hydrogen storage. There are a few reviews on the hydrogen storage on metal hydrides. The interested readers can consult them [68–70].

6.5.3 ZEOLITES

Zeolites are alumino silicate minerals that are highly crystalline and microporous. These have the capacity to enclose non-polar gases like H_2 because to their cage and tunnel architectures. Through an adsorption mechanism that involves hydrogen being driven into the pores at low temperature, and pressure hydrogen is physisorbed on zeolite surface of pores materials [71]. Since molecular hydrogen interacts with micropores inside surfaces, pore volume, BET surface area, and working environments like temperature and pressure all play a role in how much hydrogen may be stored in a porous material, similar to other porous materials [72]. According to research, one of the factors controlling this capacity, particularly under high pressure, is the channel diameter. A good material in this situation would have a lot of pores and a channel diameter close to the hydrogen molecule's kinetic diameter (dH = 2.89).

Not much has been done on hydrogen storage by zeolites. However, some papers have been found in the literature and discussed herein. Bae et al. [73] prepared zeolites based on organic ion exchangers. The authors reported maximum hydrogen sorption by organic ion-exchanged zeolite-Y (OZ) of 0.15–0.34 wt% at 298 K and 10 MPa. Chung [74] studied the hydrogen sorption on microporous zeolites with different pore sizes. The authors reported maximum (0.4 wt%) sorption of hydrogen on ultra-stable zeolite at 50 bar. Furthermore, hydrogen sorption increased with an increase in Si/Al ratio, which was attained by dealumination. Yuksel et al. [75] described hydrogen sorption on zeolite clusters. The parameters determined were adsorption energy, enthalpy, electronegativity, chemical potential, chemical hardness, HOMO energy, and LUMO energy. As per the results presented in this article, it may be concluded that OFF zeolite may be a good cryo-adsorbent for hydrogen storage. Erdogan et al. [76] prepared zeolite/carbon composites with 1,694 m^2/g surface area and 0.780 cm^3/g micropore volume for hydrogen storage. The hydrogen storage capacity was found to be 1.3 wt%, which is 1.47 times for AC and 4.16 times for zeolite.

Besides, the maximum work carried out by modeling studies on hydrogen sorption is on zeolites. Ozturk [77] carried out a modeling study for hydrogen storage on Chabazite-type zeolite with lithium. The authors used density functional theory (DFT) calculations to study the effect of surface area, charge transfers, and electrostatic potentials. The hydrogen storage capacity increased by 6 and 3 times. Wang et al. [78] discussed the zeolites' molecular insight for hydrogen storage. For this purpose, a series of modeling was carried out at 30–70 MPa pressure and 250–270 K temperature. The results showed that both pressure and temperature are crucial parameters for hydrogen storage. The variation of temperature and pressure resulted in various combinations of hydrogen storage on zeolites. Alizadeh et al. [79] used machine learning methods to predict hydrogen storage on several zeolites (28 zeolites). The authors reported cascade feed-forward neural networks as the best model based on the results obtained, with average relative deviation (7.24%), mean absolute error (0.041), root mean squared error (0.058), and regression coefficient (0.99429). The final results presented showed a good association between hydrogen uptake and zeolite capacity. Hai et al. [80] carried out modeling of hydrogen storage on zeolite to associate pressure, pores, and temperature conditions. The model predicted massive experimental data with 0.99875 regression coefficient and 6.43% absolute average relative deviation, indicating good applicability.

The Canonical Monte Carlo is the most used model for hydrogen sorption on zeolites. Song and Noo [81] carried out modeling of hydrogen sorption on organic zeolites. The model used was Grand Canonical Monte Carlo (GCMC). The effects of volume, cations surface area, pressure, temperature, and pre-sorbed benzene on hydrogen sorption were carried out. The hydrogen sorption was found to be affected by the above-mentioned factors. Prasanth et al. [82] used zeolites exchanged with rhodium and nickel for hydrogen adsorption at 77.4 K and 1.0 bar. The GCMC simulations were used for the adsorption of hydrogen in rhodium and nickel-exchanged zeolites at 77.4, 303, and 333 K. The modeling data was compared with experimental results. Rahmati and Modarress [83] also used the GCMC model for hydrogen sorption on zeolite at 250–325 K and 0–10 kbar. The effects of pressure, temperature, and pores on hydrogen adsorption were studied. The results indicated that hydrogen molecules sorbed at high pressure are determined by pore diameter. Similarly, Deniz [84] computed zeolites-templated carbons for hydrogen storage. The model used was GCMC and the hydrogen storage was associated with specific surface area, pore volume, and void fractions. The maximum hydrogen uptake achieved was 9.23 wt% at 100 bar and 77 K.

6.5.4 NANOMATERIALS

The efficient and practical use of hydrogen as a clean energy carrier is made possible by research and development in the area of hydrogen storage using nanomaterials. Due to its high energy density per mass, hydrogen is a prime choice for use in fuel cells, portable power supplies, and environmentally friendly transportation. However, there have been considerable obstacles to its wider adoption, including its low energy density by volume and the difficulties connected with safe and effective storage. Nanomaterials have special qualities that can help with some of the problems associated with hydrogen storage. The adsorption and desorption of hydrogen molecules can be improved by their high surface area, variable pore diameters, and enhanced surface interactions. Hydrogen storage can be made more effective and cost-effective through physisorption in nanoporous materials. Molecular hydrogen adsorbs rapidly and reversibly at big density on the substances' inner surfaces including ACs zeolites, and MOFs at small temperatures (e.g., 77 K) and moderate pressures (below 100 bar). The nano-materials may be of different materials like hydrides, MOFs, carbon nanotubes (CNTs), ACs, zeolites, silica graphene, etc.

There have been suggestions for hydrogen transporters made of carbon nanostructures (such as nanotubes and carbon buckyballs). However, the hydrogen contents range from 3.0 to 7.0 wt% at 77 K, which is far lower than the US DOE's target of 6 wt% at almost ambient temperatures. CNTs have been doped with MgH_2 in order

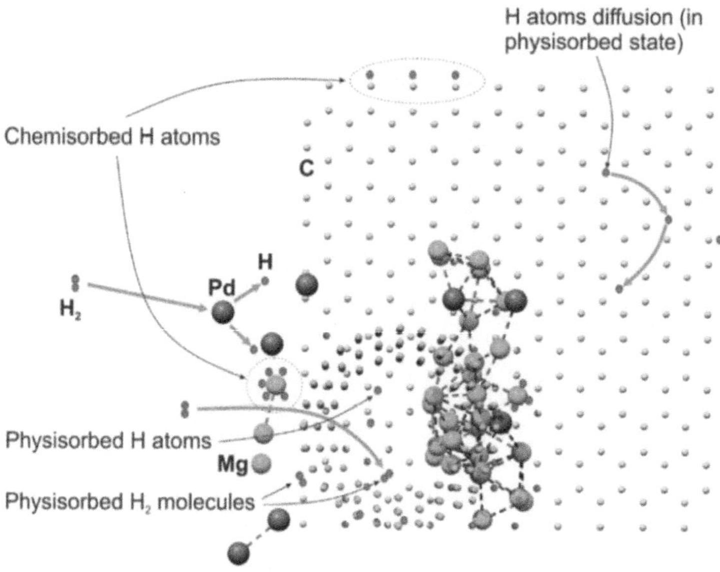

FIGURE 6.2 A schematic representation of hydrogen sorption in G/f-MWCNT@PdMg sample [reproduced with permission from ref. 94].

to realize carbon materials as efficient hydrogen storage technologies [85]. Although the metal hydride has a theoretical storage size (7.6 wt%) that exceeds the 6 wt% standard set by the US DOE. Its high release temperature limits its use in real-world applications. The CNTs in the MgH_2 lattice build fast diffusion channels as part of the hypothesized mechanism. CNTs have been used to prepare composites for the storage of hydrogen. Many researchers tried to add CNTs in different materials and prepare nano-composites and tried to store the hydrogen. The most important include the nano-composites of CNTs with Mg [86], Mg-Fe [87], $Mg-Nb_2O5$ [88], MgH_2 [89], CoB [90], Li [91], FeB [92], and PdMg [93] (Figure 6.2).

Other carbon-containing nanomaterials have also been investigated for hydrogen store at this facility, including fullerene compounds. The double-bonded carbons of fullerene molecules can be hydrogenated due to their C_{60} close-caged structure, which results in a theoretical $C_{60}H_{60}$ isomer with a hydrogen concentration of 7.7 wt%. However, these devices have a high (600°C) release temperature. The literature is scanty on the use of native fullerene for hydrogen storage. Most of the papers describe the nano-composite of fullerene with various doping agents. The most important are alkali and alkaline-earth metals [94–97], and transition metals [98–102].

Shi et al. [103] described 3D fullerene-nanotube interconnected material for hydrogen storage. The authors performed the GCMC simulations and reported 7.7–9.2 wt% hydrogen storage capacity at 298 and 233 K. Gaboardi et al. [104] described an extension of hydrogen storage in fullerenes by performing a spin relaxation experiment. A linear relationship was found between C_{60} hydrogenation level and muonium adduct hyperfine frequency of radicals, confirming the applicability of this technique in the area of hydrogen storage. Durbin et al. [105] used DFT to calculate hydrogen storage in fullerenes. The authors have analyzed several carbon nano-systems and proposed probable materials for H_2 storage for vehicles. Huang et al. [106] calculated $C_{20}\text{-}4C_5$ and $C_{20}\text{-}4B_3N_2$ terminated chain lengths by using DFT. The results presented indicated 5.7 and 5.6 wt% densities of hydrogen, conforming that both chains terminated are the best materials. Other authors have also used DFT studies to find out the hydrogen storage capacities of fullerene and its composite materials [107–110].

6.5.5 GRAPHENE AND ITS DERIVATIVES

Hydrogen can be stored in graphene and its derivatives. Naturally, graphene is chemically inert to adsorb hydrogen but its doped compounds may be used as a good sorbent for hydrogen storage. Graphene has been dropped with alkali metals, transition metals, etc. There are two ways that hydrogen can adsorb on graphene: either through chemisorption, which involves creating a chemical bond with carbon atoms, or through physisorption, which involves interacting through Van der Waals forces. The majority of the time, hydrogen molecules undergo physisorption. Theoretically, H_2 binding energy was estimated to be between 0.0 and 0.06 eV [111]. This wide range of values is due to how elusive and challenging it is to model London dispersion forces. Despite this, it is obvious that molecular hydrogen has very poor binding and needs high pressures and low temperatures to maintain good storage stability because of dissociative adsorption of H_2. The chemisorption of atomic hydrogen, however, is a rather advantageous process. In fact, the generally established values for H-binding energy and chemi-sorption barriers are 0.70 eV and 0.30 eV, respectively. The highest GD that graphene can achieve through chemisorption is 8.3% (= 1/12).

Graphene may store hydrogen effectively and hydrogen may interact with graphene network [112,113]. Al-Hamdani et al. [114] decorated graphene sheets with alkali and alkaline-earth metals hydrogen adsorption. The authors also studied the mechanisms of hydrogen binding with graphene materials and reported three adsorption mechanisms: Waals physisorption, Kubas adsorption, and metal adatom-facilitated polarization. The authors described Kubas sorption as distressed by an outside electric field, giving a way to optimize hydrogen sorption. Sunnardianto et al. [115] presented a way to modify graphene for better hydrogen sorption by using a combined molecular dynamics (MD) and DFT approach. The authors studied pristine graphene and two-dimensional graphene structures with hydrogenated vacancy (V222) and reported more reversible adsorption and desorption processes for V222. Furthermore, the authors reported V222 as sturdier giving more chance to be used for further research. In a similar study by Ao and Peeters [116] it was reported that both sides of the aluminum-sorbed graphene layer stored hydrogen up to 13.79 wt% with average sorption energy −0.193 eV/H2. It is important to mention that the power of interactions between two hydrogen adatoms and graphene magnetic properties is intensely reliant on the residence of two adatoms on the graphene sub-lattices. Hydrogen adatoms create lattice distortion and electron localization in graphene. This facilitates the attractive interactions between two H adatoms [117] A similar study was presented by Ivanovskaya et al. [118] describing an ab initio study of hydrogen sorption on graphene. The authors reported the hydrogen chemisorption energy barrier as independent of the validated technique and size of the system. A similar study was presented by Zhang et al. [119] with aluminum-modified silicon-doped single-layer graphene after hydrogen sorption by first-principles calculations.

Wu et al. [120] reported hydrogen isotope sorption by applying incident energy. The authors described that heavier isotopes exhibited higher sorption between 1.0 and 50.0 eV. Also, the sorption rates were higher with a smaller angle of incident angle of the impacting atoms (0.5–5 eV). Murata and co-workers [121] tried to optimize hydrogen sorption on graphene via gate voltage. The authors used scanning tunneling microscopy, electrical transport measurements, and functional theory calculations to achieve the task. These authors reported more hydrogen sorption with negative gate voltage (p-type doping) in comparison to positive gate voltage (n-type doping). Rostami et al. [122] prepared graphene-C_3N_4 samples and used them for the sorption of hydrogen. The maximal hydrogen storing capabilities at 22 bar and 296 K were 1.27 and 1.06 mmol/g for graphene-C_3N_4 and graphite oxide samples. The isosteric heat of hydrogen sorption on graphene oxide differs from 8.60 kJ/mol (at small hydrogen uptake) to 4.30 kJ/mol. The results showed that the interactions among molecules and tri-s-triazine units in graphene-C_3N_4 structure were more robust than in graphene oxide. Chen et al. [123] prepared Ni/Pd co-modified graphene materials with uniform sizes. As per the authors, palladium electronic changed and the center of the D-band relocated down, which encouraged the sorption of hydrogen. NiPd-rGO-180 showed the highest hydrogen store volume of 2.65 wt% at room temperature and 4MPa). The determined hydrogen sorption energies of Ni_2Pd_2-rGO were −20 to −0.60 eV. The graphene chemisorption can be enhanced by changing its chemical structure, especially the formation of hydrogen bonding groups and graphene curvature. Besides, the 3D structure of graphene provides a good platform for hydrogen storage. There are some reviews available on graphene and its nano-composites for hydrogen storage. These reviews describe various aspects of hydrogen storage [124–128]. Interested readers should read these articles in detail.

6.5.6   Liquid Organic Compounds Hydrogen Storage

Recently, the liquid organic hydrogen carrier (LOHC) technique has achieved a great reputation for stable and effective hydrogen storage and transport. This is because of safe and economical storage for a long time. Organic substances known as LOHCs can chemically absorb and release hydrogen. So, LOHCs can serve as hydrogen storage devices. In theory, hydrogen can be absorbed by any unsaturated chemical (organic molecules having C-C double or triple bonds) during hydrogenation [129]. Generally, organic hydrogen storage liquids are obtained from fossil fuels by refining procedures. Biomass is also used to make LOHCs owing to its unique carbon-balance features and the possibility to make aromatic and nitrogen compounds.

Dehydrated form of LOHC (an unsaturated), typically an aromatic molecule, undergoes a hydrogenation process with the hydrogen to absorb hydrogen. The hydrogenation is an exothermic process that takes place in the presence of a catalyst at extreme pressures (about 30–50 bar) and temperatures of about 150°C–200°C. This results in the formation of the equivalent saturated chemical, which can then be transported or stored at room temperature. To release the hydrogen from the LOHC once more, the now-hydrogenated, hydrogen surplus form of the LOHC is dehydrogenated. This endothermic reaction also occurs in the presence of a catalyst at high temperatures (250°C–320°C). The hydrogen may need to be cleaned of LOHC steam before it can be utilized. In order to maximize efficiency, cold material flow made up of hydrogen surplus LOHC entering the release unit should receive the heat from the hot material flow leaving it to reduce the energy needed to pre-heat it before the reaction. In particular, it is theoretically possible to utilize the heat produced by hydrogenation reaction at hydrogen sorption for heating or as process heat [130].

Imamura [131] reported an improved way of hydrogen storage by modifying the organic hydrogen storage molecules. This author reported that metal vaporization into an organic matrix at 77 K was an effective way to increase hydrogen storage. Shiraz et al. [132] stored hydrogen in LOHCs after electrochemical production by change of a donor proton into a hydrogen atom involving covalent bonds with LOHC (R) through a proton-coupled electron transfer (PCET) reaction. 9-Fluorenone/fluorenol (Fnone/Fnol) change was used as a model PCET reaction. Yu et al. [133] reported the preparation of palladium-doped N-ethylcarbazole (NEC) for hydrogen storage. As per the authors, lattice hydrogen-bonding sites made hydrogen storage greatly pretty for hydrogen. Gambini et al. [134] used control of the hydrogen flow rate in a LOHC batch reactor to increase hydrogen storage via temperature and pressure. By controlling temperature, the authors reported efficiency higher than 90%. Verevkin et al. [135] used di-phenyl ether derivatives for hydrogen storage. These derivatives were found good candidates for good hydrogen storage with low volatilities.

Heublein et al. [136] developed the design of a dehydrogenation reactor design via simulation for hydrogen storage in organic liquid carriers (NEC). The authors described tubular and vertical reactors for good efficiency and stability. Geiling et al. [137] used dibenzyltoluene (DBT)/perhydro-DBT for hydrogenation and dehydrogenation in a reactor based on PEM technology. The authors reported good performance without any serious deprivation. Verevkin et al. [138] carried out experimental and simulation studies on thermodynamics for hydrogen storage on pyrazine derivatives. The results indicated good hydrogenation/dehydrogenation of hydrogen storage. Byun et al. [139] described the economic, environmental, and technical aspects of hydrogen storage on LOHC. The authors used toluene (TOL)-methylcyclohexane (MCH), DBT, NEC-perhydro-NEC (12H-NEC), CO_2–methanol (MeOH), and perhydro-DBT (18H-DBT) molecules. The authors used Aspen Plus® for the simulation study and a good amount of hydrogen stored was 7.20 $kmolH_2.h^{-1}$ on DBT-18H-DBT. Shin et al. [140] carried out MD simulations, Conductor-like Screening Model, and DFT studies for hydrogen storage using N-phenylcarbazole, N-acetylcarbazole, 4-methyl-4H-benzocarbazole, and N-benzoylcarbazole compounds. The authors reported a theoretical value of more than 6 wt% hydrogen storage. Cao et al. [141] used artificial intelligence (AI) to compare the performance of MOFs. AI included hybrid neuro-fuzzy systems, artificial neural networks, and support vector machines as estimators. The results showed a good correlation with the literature data with 5.34% relative absolute deviation, 0.059 mean squared error, and 0.9946 coefficient of determination. Vostrikov et al. [142] used 1-alkyl-indoles as suitable LOHC for hydrogen storage. The experimental results were compared with those obtained by theoretical quantum chemical methods. The authors reported quite good hydrogenation and dehydrogenation of hydrogen on these compounds. The other researchers also attempted to develop composite/doped LOHC for hydrogen storage. Shirvani et al. [143] prepared 2-D Mo2C as a promotor for hydrogen storage and release from DBT and perhydro-DBT (H18-DBT). The addition of Pt/Al_2O_3 catalyst increased hydrogenation and dehydrogenation processes. Ozturk et al. [144] carried out experimental and modeling studies of hydrogen storage on MOF. The MOF developed were copper and zinc metals with benzene 1,3,5 tricarboxylic acid, 1,10 phenantroline, and benzene 1,3,5 tricarboxylic acid. The authors reported uptakes of Zn(II) and Cu(II) as 1.383 wt.% (1.187 wt.%) and 2.652 wt.% (sim. 2.434 wt.%) at 1 bar pressure and 77 K.

Overall, hydrogen hydrogenation and dehydrogenation on LOHC are thought to be the primary issues that lower the storage cycle's overall efficiency. There are several reviews and interested readers should consult these for further details [145–153]. These articles describe hydrogen storage applications, organic molecules chemisorption prospective, MOFs as future, thermodynamics, composite/doped LOHCs, and other optimizing parameters for maximum hydrogen storage.

6.6 COMPARISON OF STORAGE METHODS

A comparison of hydrogen storage is provided by Xu et al. [154] and the authors reported that various methods have different applications depending on the requirement (Figure 6.3). In spite of all these advantages and disadvantages, the choice of hydrogen storage method depends on the requirement and other controlling factors like storage duration, energy density, energy efficiency, release possibility, safety, cost, etc. However, there is no method perfect and economic; specially for motor vehicles and aircrafts. More and more research is needed in this important area of hydrogen storage.

Some advantages and disadvantages of different methods are summarized below.

6.6.1 COMPRESSED HYDROGEN GAS

- **Advantages:** Simple method, and low energy consumption for storage.
- **Disadvantages:** Needs high-pressure (typically 350–700 bar) tanks, big-size store tanks, energy-intensive compression, costly process, and possible safety concerns.

6.6.2 LIQUID HYDROGEN (LH$_2$)

- **Advantages:** High-density hydrogen, and long-duration storage.
- **Disadvantages:** High energy required to bring low temperature (around −253°C or 20 K), costly tanks, big-size store tanks, and complex protection requirements.

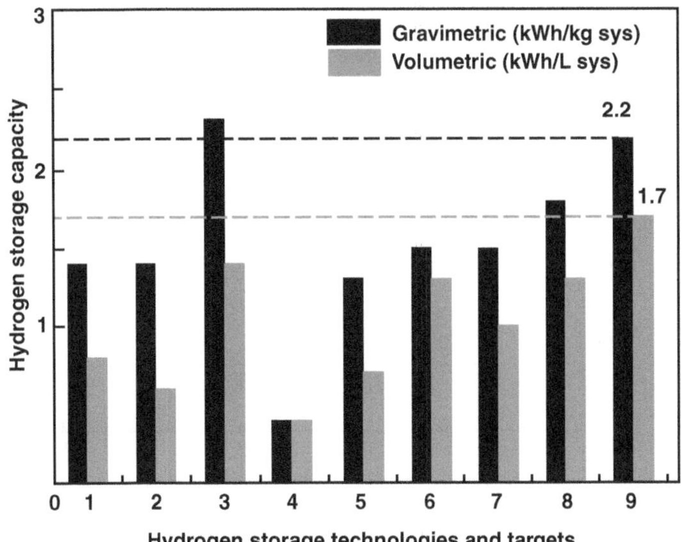

FIGURE 6.3 Predictable performance of hydrogen storage capacity comparison. 1: 700 bar compacted type IV; 2: 300 bar compacted type IV; 3: 500 bar cryo-compacted; 4: metal hydrideNaAlH$_4$/Ti; 5: sorbent MOF-5,100 bar; 6: chemical store, AB liquid; 7: 2020 goal magnitudes; 8: 2025 goal magnitudes; 9: final goal values [reproduced with permission from ref. 154].

6.6.3 Underground hydrogen storage

- **Advantages:** Large-scale storage, reduced footprint, safety, thermal management, preservation of purity, and energy efficiency.
- **Disadvantages:** Technical challenges, geological constraints, monitoring and maintenance, regulatory approval, scale-up challenges, and limited locations.

6.6.4 Adsorption storage

- **Advantages:** Modest pressures, good hydrogen storage, low heat management need, and a choice of developing new materials including nanomaterials.
- **Disadvantages:** Adsorption saturated easily, limited storage ability, cooling during adsorption, low hydrogen uptake, and release kinetics.

6.7 ECONOMY OF DIFFERENT HYDROGEN STORAGE METHODS

It is vital to keep in mind that the economics of hydrogen storage systems depend on several variables, including the amount of storage needed, the state of technology, needs of the infrastructure, safety concerns, and the price of hydrogen production.

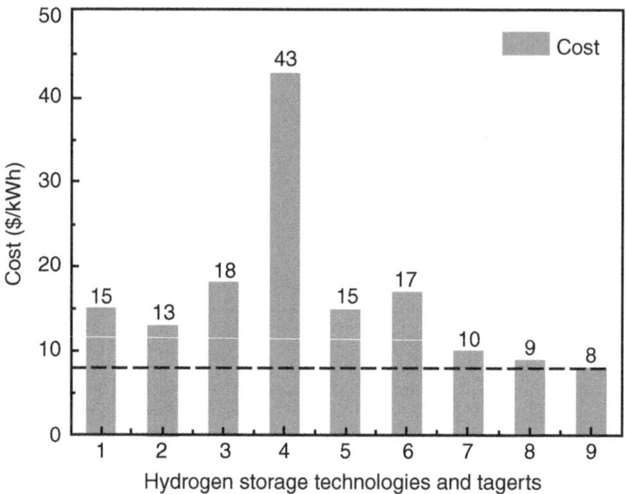

FIGURE 6.4 Predicted comparison of cost, 500 k units/yr. 1: 700 bar compacted type IV; 2: 300 bar compacted type IV; 3: 500 bar cryo-compacted; 4: metal hydride NaAlH$_4$/ Ti; 5: adsorbent MOF-5,100 bar; 6: chemical store, AB liquid; 7: 2020 goal magnitudes; 8: 2025 goal magnitudes; 9: final goal magnitudes [reproduced with permission from ref. 154].

The gravimetric and volumetric densities, store release and uptake kinetics, operational safety, and cost all contribute to the performance of surface storage techniques. Subterranean geologic media, such as salt caverns, exhausted hydrocarbon basins, aquifers, and hard rock caverns, can be used to store large amounts of seasonal hydrogen. The targeted storage cycles, capacities, and purity of the hydrogen stored are all affected by the nature of geo-structures. The surface storage techniques make it easier to store hydrogen for fixed and mobile applications that are often smaller in capacity and call for quick storage cycles. Hydrogen is physically stored by being contained in vessels in all of its various phases, including compressed gaseous, cryogenic, and cryo-compressed forms. By absorbing or adsorbing hydrogen utilizing solid-state materials, hydrogen can be stored materially.

The hydrogen store materials research is vast with many publications. The maximum papers are from China, the EU, and the USA, followed by Japan. According to a literature survey from 2000 to 2023, the research in hydrogen storage was on a boom from 2000 to 2010 and later became sleepy with a plateau in 2015. MOFs and simple hydrides were the material classes that received the most investigation. The commercialization of existing physio-sorption technology is still a long way off. The small samples, fewer than 100 g, are used in the experimental experiments [155]. The technologies mentioned typically call for high pressures and/or low temperatures. Because of this, these approaches are not regarded as unique technologies in and of themselves at their current state of the art, but rather as a kind of useful supplement to existing compression and liquefaction procedures. Since there is no activation energy and very little interaction energy, physiological sorption processes are reversible. Hydrogen is physisorbed on surface of pores in substances including porous carbons, MOFs, zeolites, organic polymers, and clathrates. As a result, many physio-sorption-based materials have large store capabilities at liquid nitrogen temperatures and pressures, but their capabilities are drastically reduced at room temperature and pressure. Each storage technology's technical insights are offered together with suggestions and pertinent application domains. No storage technology can be said to be the best fit for all uses under ideal circumstances, and each technology needs significant development before it can be used to store energy. The cost of storing hydrogen is anticipated to go down as technology develops and economies of scale are reached, making hydrogen a more viable and competitive energy storage option. A cost-wise comparison is provided by Xu et al. [154] and the authors reported that various methods have different economies for different purposes. Figure 6.4 indicates the comparison of cost, predicted to 500 k units/yr of hydrogen storage. Interested readers should consult the latest review articles available in the literature [154,156,157].

6.8 PORTABILITY OF HYDROGEN STORAGE ON BOARD

The application of hydrogen on board is needed in hydrogen fuel cell vehicles, hydrogen internal combustion engines, hydrogen fuel cells for aviation, hydrogen-powered ships, space travel, emergency backup power, portable power generators, forklifts and warehouse equipment, etc. On-board hydrogen storage is the real demand of today and the future. The highly accessible hydrogen capabilities in terms of volume and weight, safety, hydrogen release, and recharge at moderate temperatures (and costs) are the major requirements for an ideal on-board tank. It is important to mention that it is difficult to achieve such type of tank on board. The compressed hydrogen containers at 350.0 bar (50,000 psi) and 700.0 bar (10,000.0 psi) are utilized in vehicles, based on type IV carbon-composite techniques. However, the liquid and compressed hydrogen store choices are good for on-board applications. The cryogenic LH_2 has restricted uses. Nowadays, passenger cars, light trucks, etc. depend on compressed hydrogen tanks. The metal hydride-based containers are safe and comparatively compact but their high weight is the major limitation. Still, physical hydrogen storage technologies are more suitable for motor cars, trucks, trains, and aircrafts while material-based storage systems may be suitable for light passenger cars until a good thermal management design is developed. LOHCs may be another attractive on-board hydrogen carrier in the future. Not much is developed in this direction and challenges exist such as operation conditions, the cost, and the refueling period. Slow kinetics during hydrogen cycling and high thermodynamic stability affect its practical application on board. These are the reasons for high pressure and temperature in operation.

Much research has been done on exploring the required hydrogen storage on board, but still the compressed or LH_2 tanks are being used. The most important research done includes the safe, compact, reliable, economical, and efficient storage method on board. Many theoretical calculations and models are presented but the real experimental results need more to get the reality of on-board hydrogen storage. However, we still lack hydrogen storage on board as per the targets of the US DOE. The storage must have at least 5–10 kg to achieve a long-range journey (~500 km). Mercedes-Benz, Hyundai, Toyota, and Honda offered hydrogen-based cars in 2021 globally. Briefly, light-weight and small storage systems are needed due to the weight and size constraints in vehicles, which is a challenge for researchers. The researchers and engineers are continuously working to develop creative storage methods that balance these features. Hydrogen may become more portable and more widely used as an energy carrier as technology develops and hydrogen storage systems get better.

6.9 CONCLUSION

The perfect hydrogen storage device should be inexpensive, strong, with a huge quantity of hydrogen storage capability, a small recharge time, and workable in room conditions. Concerning all the methods discussed in this chapter all of them have pros and cons. The main problems with these methods are expense, low amount of hydrogen storage, difficulty to transport, and requirement of energy to get back hydrogen for any use. However, the sorption of hydrogen on solid materials may be the future of hydrogen storage. Among many methods graphene and its derivatives are gaining importance in this area of research owing to their unique features, especially the large surface area and light-weight. A number of methods for hydrogen storage in graphene structures are being currently studied by theoretical approaches. The theoretical studies suggested that graphene derivatives can adsorb by 9.0% mass ratio of hydrogen. Still, experimental verification is needed from the laboratory and then industries. To achieve all these in reality, there is a great need to develop more graphene-based materials with huge strength, heat and electrical conductivity, flexibility, and manageability. Also, the use of a novel 3D graphene structure permits the limitations of 2D graphene materials. A multilayer graphene storage arrangement with the controlled alteration of the curvature for storage and release of hydrogen can work at room conditions with fast kinetics. The light-weight, sturdiness, and excellent heat and electricity conductivities made graphene the ideal material for hydrogen storage, especially in transportation. Besides, the LOHC area is gaining importance in hydrogen storage. These organic compounds may be the future of hydrogen storage. The sandwiched structures with organic compounds and graphene may be a good future for hydrogen storage.

REFERENCES

1. V. Tozzini, V. Pellegrini, Prospects for hydrogen storage in graphene, *Phys. Chem. Chem. Phys.*, 15, 80–89 (2013).
2. U. Eberle, B. Mueller, R. von Helmolt, Fuel cell electric vehicles & hydrogen infrastructure: Status 2012, *Energy Environ. Sci.*, 5, 8780–8798 (2012).
3. H. Barthélémy, M. Weber, F. Barbier, Hydrogen storage: Recent improvements & industrial perspectives, *Int. J. Hydrogen Energy*, 42, 7254–7262 (2017).
4. M. Li, Y. Bai, C. Zhang, Y. Song, S. Jiang, D. Grouset, M. Zhang, Review on the research of hydrogen storage system fast refueling in fuel cell vehicle, *Int. J. Hydrogen Energy*, 44, 10677–10693 (2019).
5. R. Moradi, K.M. Groth, Hydrogen storage & delivery: Review of the state of the art technologies & risk & reliability analysis, *Int. J. Hydrogen Energy* 44 12254–12269 (2019).

6. H. Barthélémy, Hydrogen storage – Industrial prospectives, *Int. J. Hydrogen Energy*, 37, 17364–17372 (2012).
7. H.T. Hwang, A. Varma, Hydrogen storage for fuel cell vehicles, *Curr. Opin. Chem. Eng.*, 5, 42–48 (2014).
8. M.S. Sadaghiani, M. Mehrpooya, Introducing & energy analysis of a novel cryogenic hydrogen liquefaction process configuration, *Int. J. Hydrogen Energy*, 42 (9), 6033–6050 (2017).
9. H.T. Hwang, A. Varma, Hydrogen storage for fuel cell vehicles, *Curr. Opin. Chem. Eng.*, 5, 42–48 (2014).
10. S. Lasher, Analyses of hydrogen storage materials & on-board systems archived 2011–09–29 at the Wayback Machine, *DOE Annu. Merit Rev.*, June 7–11 (2010).
11. R. Ahluwalia, T. Hua, J.K. Peng, S. Lasher, K. McKenney, J. Sinha, M. Gardiner, Technical assessment of cryo-compressed hydrogen storage tank systems for automotive applications, *Int. J. Hydrogen Energy*, 4171–4184 (2010).
12. R.K. Ahluwalia, J.K. Peng, H.S. Roh, T.Q. Hua, C. Houchins, B.D. James, Supercritical cryo-compressed hydrogen storage for fuel cell electric buses, *Int. J. Hydrogen Energy*, 43, 10215–10231(2018).
13. S.M. Aceves, F. Espinosa-Loza, E. Ledesma-Orozco, T.O. Ross, A.H. Weisberg, T.C. Brunner, O. Kircher, High-density automotive hydrogen storage with cryogenic capable pressure vessels, *Int. J. Hydrogen Energy*, 35, 1219–1226 (2010).
14. A. Hassanpouryouzband, E. Joonaki, K. Edlmann, R.S. Haszeldine, Offshore geological storage of hydrogen: Is this our best option to achieve net-zero? *ACS Energy Lett.*, 6(6), 2181–2186 (2021).
15. F. Chen, Z. Ma, H. Nasrabadi, B. Chen, M.Z.S. Mehana, J.V. Wijk, Capacity assessment & cost analysis of geologic storage of hydrogen: A case study in Intermountain-West Region USA, *Int. J. Hydrogen Energy*, 48, 9008–9022 (2023).
16. N. Lin, L. Xu, L.G. Moscardelli, Market-based asset valuation of hydrogen geological storage, *Int. J. Hydrogen Energy*, 49, 114–129 (2024).
17. M. Ghaedi, P.Ø. &ersen, R. Gholami, Maximum column height & optimum storage depth for geological storage of hydrogen, *Int. J. Hydrogen Energy*, 50, 291–304 (2023).
18. R. Tarkowski, B. Uliasz-Misiak, P. Tarkowski, Storage of hydrogen, natural gas, & carbon dioxide – Geological & legal conditions, *Int. J. Hydrogen Energy*, 38, 20010–20022 (2021).
19. A. Lemieux, A. Shkarupin, K. Sharp, Geologic feasibility of underground hydrogen storage in Canada, *Int. J. Hydrogen Energy*, 45, 32243–32259 (2020).
20. J. Mouli-Castillo, N. Heinemann, K. Edlmann, Mapping geological hydrogen storage capacity & regional heating demands: An applied UK case study, *Appl. Energy*, 283, 116348 (2021).
21. J.D.O. Williams, J.P. Williamson, D. Parkes, D.J. Evans, K.L. Kirk, N. Sunny, E. Hough, H. Vosper, M.C. Akhurst, Does the United Kingdom have sufficient geological storage capacity to support a hydrogen economy? Estimating the salt cavern storage potential of bedded halite formations, *J. Energy Storage*, 53, 2022, 105109.
22. R. Gholami, Hydrogen storage in geological porous media: Solubility, mineral trapping, H2S generation & salt precipitation, *J. Energy Storage*, 59, 2023, 106576.
23. R. Ershadnia, M. Singh, S. Mahmoodpour, A. Meyal, F. Moeini, S.A. Hosseini, D.M. Sturmer, M. Rasoulzadeh, Z. Dai, M.R. Soltanian, Impact of geological & operational conditions on underground hydrogen storage, *Int. J. Hydrogen Energy*, 48, 1450–1471 (2023).
24. A. Alanazi, N. Yekeen, M. Ali, M. Ali, I.S. Abu-Mahfouz, A. Keshavarz, S. Iglauer, H. Hoteit, Influence of organics & gas mixing on hydrogen/brine & methane/brine wettability using Jordanian oil shale rocks: Implications for hydrogen geological storage, *J. Energy Storage*, 62, 106865 (2023).

25. J. Ikäheimo, T.J. Lindroos, J. Kiviluoma, Impact of climate & geological storage potential on feasibility of hydrogen fuels, *Appl. Energy*, 342, 121093 (2023).

26. F. Alhamad, M. Ali, N.P. Yekeen, M. Ali, H. Hoteit, S. Iglauer, A. Keshavarz, Effect of methylene blue on wetting characteristics of quartz/H2/brine systems: Implication for hydrogen geological storage, *J. Energy Storage*, 72,108340 (2023).

27. A. Hassanpouryouzband, E. Joonaki, K. Edlmann, R.S. Haszeldine, Offshore geological storage of hydrogen: Is this our best option to achieve net-zero, *ACS Energy Lett.*, 6, 2181–2186 (2021).

28. N.S. Muhammed, B. Haq, D. Al Shehri, A. Al-Ahmed, M.M. Rahman, E. Zaman, A review on underground hydrogen storage: Insight into geological sites, influencing factors & future outlook, *Energy Rep.*, 8, 461–499 (2022).

29. T. Amirthan, M.S.A. Perera, Underground hydrogen storage in Australia: A review on the feasibility of geological sites, *Int. J. Hydrogen Energy*, 48, 4300–4328 (2023).

30. CNRS Institut Neel H2 Storage Archived 2016-03-03 at the Wayback Machine. Neel. cnrs.fr.

31. J. Graetz, J. Reilly, G. Sandrock, J. Johnson, W. Zhou, J. Wegrzyn, Aluminum hydride, AlH3, As a hydrogen storage compound (2006). doi:10.2172/899889.

32. H. Wang, M. Du, Q. Wang, Z. Li, S. Wang, Z. Gao, J.J. Derksen, Enhancement of hydrogen storage performance in shell & tube metal hydride tank for fuel cell electric forklift, *Int. J. Hydrogen Energy*, 48, 23568–23580 (2023).

33. J. Kotowicz, W. Uchman, M. Jurczyk, R. Sekrt, Evaluation of the potential for distributed generation of green hydrogen using metal-hydride storage methods, *Appl. Energy*, 344, 121269 (2023).

34. P. Di Giorgio, G. Di Ilio, E. Jannelli, F.V. Conte, Numerical analysis of an energy storage system based on a metal hydride hydrogen tank & a lithium-ion battery pack for a plug-in fuel cell electric scooter, *Int. J. Hydrogen Energy*, 48, 3552–3565 (2023).

35. H. Chang, Y.B. Tao, H. Ye, Numerical study on hydrogen & thermal storage performance of a s&wich reaction bed filled with metal hydride & thermochemical material, *Int. J. Hydrogen Energy*, 48, 20006–20019 (2023).

36. S. Jana, P. Muthukumar, Design, development & hydrogen storage performance testing of a tube bundle metal hydride reactor, *J. Energy Storage*, 63, 106936 (2023).

37. Y. Ye, J. Lu, J. Ding, W. Wang, J. Yan, Performance improvement of metal hydride hydrogen storage tanks by using phase change materials, *Appl. Energy*, 320, 119290 (2022).

38. L. Zhan, P. Zhou, X. Xiao, Z. Cao, M. Piao, Z. Li, L. Jiang, Z. Li, L. Chen, Numerical simulation & experimental validation of Ti0.95Zr0.05Mn0.9Cr0.9V0.2 alloy in a metal hydride tank for high-density hydrogen storage, *Int. J. Hydrogen Energy*, 47, 38655–38670, (2022).

39. D. Wang, Y. Wang, F. Wang, S. Zheng, S. Guan, L. Zheng, L. Wu, X. Yang, M. Lv, Z. Zhang, Optimal design of disc mini-channel metal hydride reactor with high hydrogen storage efficiency, *Appl. Energy*, 308, 118389 (2022).

40. T. Disli, S.A. Çetinkaya, M.A. Ezan, C.O. Colpan, Numerical investigations on the absorption of a metal hydride hydrogen storage tank based on various thermal management strategies, *Int. J. Hydrogen Energy*, 51, 504–522 (2023).

41. R. Elkhatib, H. Louahlia, Metal hydride cylindrical tank for energy hydrogen storage: Experimental & computational modeling investigations, *Appl. Therm. Eng.*, 230, 120756 (2023).

42. M. Afzal, N. Sharma, N. Gupta, P. Sharma, Transient simulation studies on a metal hydride based hydrogen storage reactor with longitudinal fins, *J. Energy Storage*, 51, 2022, 104426.

43. Y. Li, E. Teliz, F. Zinola, V. Díaz, Design of a AB5-metal hydride cylindrical tank for hydrogen storage, *Int. J. Hydrogen Energy*, 46, 33889–33898 (2021).

44. R. Aruna, S.T. Jaya Christa, Modeling, system identification & design of fuzzy PID controller for discharge dynamics of metal hydride hydrogen storage bed, *Int. J. Hydrogen Energy*, 45, 4703–4719 (2020).
45. Y. Luo, Q. Wang, J. Li, F. Xu, L. Sun, Y. Zou, H. Chu, B. Li, K. Zhang, Enhanced hydrogen storage/sensing of metal hydrides by nanomodification, *Mater. Today Nano*, 9, 2020, 100071.
46. H.Q. Nguyen, B. Shabani, Review of metal hydride hydrogen storage thermal management for use in the fuel cell systems, *Int. J. Hydrogen Energy*, 62, 31699–31726 (2021).
47. Z. Dong, Y. Wang, H. Wu, X. Zhang, Y. Sun, Y. Li, J. Chang, Z. He, J. Hong, A design methodology of large-scale metal hydride reactor based on schematization for hydrogen storage, *J. Energy Storage*, 49, 104047 (2022).
48. G.R. de Almeida Neto, F.H. Matheus, C.A.G. Beatrice, D.R. Leiva, L.A. Pessan, Fundamentals & recent advances in polymer composites with hydride-forming metals for hydrogen storage applications, *Int. J. Hydrogen Energy*, 47, 34139–34164 (2022).
49. Y. Liu, W. Zhang, X. Zhang, L. Yang, Z. Huang, F. Fang, W. Sun, M. Gao, H. Pan, Nanostructured light metal hydride: Fabrication strategies & hydrogen storage performance, *Renew. Sustain. Energy Rev.*, 184, 113560 (2023).
50. N. Klopčič, I. Grimmer, F. Winkler, M. Sartory, A. Trattner, A review on metal hydride materials for hydrogen storage, *J. Energy Storage*, 72, 108456 (2023).
51. F.J. Desai, M.N. Uddin, M.M. Rahman, R. Asmatulu, A critical review on improving hydrogen storage properties of metal hydride via nanostructuring & integrating carbonaceous materials, *Int. J. Hydrogen Energy*, 48, 29256–29294 (2023).
52. N. Klopčič, I. Grimmer, F. Winkler, M. Sartory, A. Trattner, A review on metal hydride materials for hydrogen storage, *J. Energy Storage*, 72, 108456 (2023).
53. M. Sevilla, R. Mokaya, Energy storage applications of activated carbons: Supercapacitors & hydrogen storage, *Energy Environ. Sci.*, 7 (4), 1250–1280 (2014).
54. L.S. Blankenship, N. Balahmar, R. Mokaya, Oxygen-rich microporous carbons with exceptional hydrogen storage capacity, *Nat. Commun.*, 8, 1545 (2017).
55. T.S. Blankenship, R. Mokaya, Cigarette butt-derived carbons have ultra-high surface area & unprecedented hydrogen storage capacity, *Energy Environ. Sci.*, 10 (12), 2552–2562 (2017).
56. R. Pedicini, S. Maisano, V. Chiodo, G. Conte, A. Policicchi, R.G. Agostino, Posidonia Oceanica & Wood chips activated carbon as interesting materials for hydrogen storage, *Int. J. Hydrogen Energy*, 45, 14038–14047 (2020).
57. M. Doğan, P. Sabaz, Z. Bicil, B. Koçer Kizilduman, Y. Turhan, Activated carbon synthesis from tangerine peel & its use in hydrogen storage, *J. Energy Inst.*, 93, 2176–2185 (2020).
58. M. Pirsaheb, T. Gholami, E.A. Dawi, H.S. Majdi, F.S. Hashim, H. Seifi, M. Salavati-Niasari, Green synthesis & characterization of SnO_2, CuO, Fe_2O_3/activated carbon nanocomposites & their application in electrochemical hydrogen storage, *Int. J. Hydrogen Energy*, 48, 23594–23606 (2023).
59. X. Li, H. Tian, S. Yan, H. Shi, J. Wu, Y. Sun, Y. Xing, H. Bai, H. Zhang, Micropores enriched ultra-high specific surface area activated carbon derived from waste peanut shells boosting the performance of hydrogen storage, *Int. J. Hydrogen Energy*, 50, 324–336 (2023).
60. E. Çetingürbüz, A. Turkyilmaz, Production of activated carbon by lithium activation & determination of hydrogen storage capacity, *Ind. Crops Prod.*, 203, 117171(2023).
61. N.M. Musyoka, M. Wdowin, K.M. Rambau, W. Franus, R. Panek, J. Madej, D. Czarna-Juszkiewicz, Synthesis of activated carbon from high-carbon coal fly ash & its hydrogen storage application, *Renewable Energy*, 155, 1264–1271 (2020).

62. N. Kostoglou, C. Koczwara, S. Stock, C. Tampaxis, G. Charalambopoulou, T. Steriotis, O. Paris, C. Rebholz, C. Mitterer, Nanoporous polymer-derived activated carbon for hydrogen adsorption & electrochemical energy storage, *Chem. Eng. J.*, 427, 131730 (2022).

63. L. Luo, Y. Zhou, W. Yan, L. Luo, J. Deng, M. Fan, W. Zhao, In-situ one-step synthesis of activated Carbon@MIL-101 (Cr) composites for hydrogen storage, *Int. J. Hydrogen Energy*, 47, 39563–39571 (2022).

64. M. Wu, Q. Zheng, T. Sun, X. Zhang, Analysis of heat conducting enhancement measures on the composite for hydrogen storage by incorporation of activated carbon with MOFs, *Int. J. Hydrogen Energy*, 48, 3994–4005 (2023).

65. A. Morandé, P. Lillo, E. Blanco, C. Pazo, A.B. Dongil, X. Zarate, M. Saavedra-Torres, E. Schott, R. Canales, A. Videla, N. Escalona, Modification of a commercial activated carbon with nitrogen & boron: Hydrogen storage application, *J. Energy Storage*, 64, 107193 (2023).

66. A. Chibani, G. Mecheri, A. Dehane, S. Merouani, I. Ferhoune, Performance improvement of adsorptive hydrogen storage on activated carbon: Effects of phase change material & inconstant mass flow rate, *J. Energy Storage*, 56, 105930 (2022).

67. Y. Li, H. Liu, C. Yang, M. Zhu, T. Chen, The activation & hydrogen storage characteristics of the cup-stacked carbon nanotubes, *Diamond Relat. Mater.*, 100, 107567 (2019).

68. P. Manasa, S. Sambasivam, F. Ran, Recent progress on biomass waste derived activated carbon electrode materials for supercapacitors applications-A review, *J. Energy Storage*, 54, 105290 (2022).

69. S. Mandal, J. Hu, S.Q. Shi, A comprehensive review of hybrid supercapacitor from transition metal & industrial crop-based activated carbon for energy storage applications, *Mater. Today Commun.*, 34, 105207 (2023).

70. P. Ndagijimana, H. Rong, P. Ndokoye, J.P. Mwizerwa, F. Nkinahamira, S. Luo, D. Guo, B. Cui, A review on activated carbon/graphene composite-based materials: Synthesis & applications, *J. Cleaner Prod.*, 417, 138006 (2023).

71. J. Dong, X. Wang, H. Xu, Q. Zhao, J. Li, Hydrogen storage in several microporous zeolites, *Int. J. Hydrogen Energy*, 32 (18), 4998–5004 (2007).

72. J. Ren, N.M. Musyoka, H.W. Langmi, M. Mathe, S. Liao, Current research trends & perspectives on materials-based hydrogen storage solutions: A critical review, *Int. J. Hydrogen Energy*, 42 (1), 289–311 (2017).

73. D. Bae, H. Park, J.S. Kim, J.b. Lee, O.Y. Kwon, K.Y. Kim, M.K. Song, K.T. No, Hydrogen adsorption in organic ion-exchanged zeolites, *J. Phys. Chem. Solids*, 69, 1152–1154 (2008).

74. K.H. Chung, High-pressure hydrogen storage on microporous zeolites with varying pore properties, *Energy*, 35, 2235–2241 (2010).

75. N. Yuksel, A. Kose, M.F. Fellah, A density functional theory study of molecular hydrogen adsorption on Mg site in OFF type zeolite cluster, *Int. J. Hydrogen Energy*, 45, 34983–34992 (2020).

76. F.O. Erdogan, C. Celik, A.C. Turkmen, A.E. Sadak, E. Cücü, Hydrogen storage behavior of zeolite/graphene, zeolite/multiwalled carbon nanotube & zeolite/green plum stones-based activated carbon composites, *J. Energy Storage*, 72, 108471 (2023).

77. Z. Ozturk, Hydrogen storage on lithium modified silica-based CHAbazite type zeolite, A computational study, *Int. J. Hydrogen Energy*, 43, 22365–22376 (2018).

78. Y. Wang, K. Yin, S. Fan, X. Lang, C. Yu, S. Wang, S. Li, The molecular insight into the "Zeolite-ice" as hydrogen storage material, *Energy*, 217, 119406 (2021).

79. S.M. Seyed Alizadeh, Z. Parhizi, A.H. Alibak, B. Vaferi, S. Hosseini, Predicting the hydrogen uptake ability of a wide range of zeolites utilizing supervised machine learning methods, *Int. J. Hydrogen Energy*, 47, 21782–21793 (2022).

80. T. Hai, F.A. Alenizi, A.H. Mohammed, B.S. Chauhan, B. Al-Qargholi, A.S.M. Metwally, M. Ullah, Machine learning-aided modeling of the hydrogen storage in zeolite-based porous media, *Int. Commun. Heat Mass Transfer*, 145, 106848 (2023).
81. M.K. Song, K.T. No, Molecular simulation of hydrogen adsorption in organic zeolite, *Catal. Today*, 120, 374–382 (2007).
82. K.P. Prasanth, R.S. Pillai, H.C. Bajaj, R.V. Jasra, H.D. Chung, T.H. Kim, S.D. Song, Adsorption of hydrogen in nickel & rhodium exchanged zeolite X, *Int. J. Hydrogen Energy*, 33, 735–745 (2008).
83. M. Rahmati, H. Modarress, Grand canonical Monte Carlo simulation of isotherm for hydrogen adsorption on nanoporous siliceous zeolites at room temperature, *Appl. Surf. Sci.*, 255, 4773–4778 (2009).
84. C.U. Deniz, Computational screening of zeolite templated carbons for hydrogen storage, *Comput. Mater. Sci.*, 202, 110950 (2022).
85. M.U. Niemann, S.S. Srinivasan, A.R. Phani, A. Kumar, D.Y. Goswami, E.K. Stefanakos, Nanomaterials for hydrogen storage applications: A review, *J. Nanomate.*, 2008, 950967 (2008).
86. B.H. Chen, C.H. Kuo, J.R. Ku, P.S. Yan, C.J. Huang, M.S. Jeng, F.H. Tsau, Highly improved with hydrogen storage capacity & fast kinetics in Mg-based nanocomposites by CNTs, *J. Alloys Compd.*, 568, 78–83 (2013).
87. G.F. de Lima Andreani, M.R.M. Triques, C.S. Kiminami, W.J. Botta, V. Roche, A.M. Jorge Jr, Characterization of hydrogen storage properties of Mg-Fe-CNT composites prepared by ball milling, hot-extrusion & severe plastic deformation methods, *Int. J. Hydrogen Energy*, 41, 23092–23098 (2016).
88. M. Gajdics, T. Spassov, V. Kovács Kis, E. Schafler, Á. Révész, Microstructural & morphological investigations on Mg-Nb2O5-CNT nanocomposites processed by high-pressure torsion for hydrogen storage applications, *Int. J. Hydrogen Energy*, 45, 7917–7928 (2020).
89. A. Ranjbar, M. Ismail, Z.P. Guo, X.B. Yu, H.K. Liu, Effects of CNTs on the hydrogen storage properties of MgH2 & MgH2-BCC composite, *Int. J. Hydrogen Energy*, 35, 7821–7826 (2010).
90. S. Gao, H. Liu, L. Xu, S. Li, X. Wang, M. Yan, Hydrogen storage properties of nano-CoB/CNTs catalyzed MgH2, *J. Alloys Compd.*, 735, 635–642 (2018).
91. C. Baykasoglu, Z. Ozturk, M. Kirca, A.T. Celebi, A. Mugan, A.C. To, Effects of lithium doping on hydrogen storage properties of heat welded r&om CNT network structures, *Int. J. Hydrogen Energy*, 41, 8246–8255 (2016).
92. S. Gao, X. Wang, H. Liu, T. He, Y. Wang, S. Li, M. Yan, Effects of nano-composites (FeB, FeB/CNTs) on hydrogen storage properties of MgH2, *J. Power Sources*, 438, 227006 (2019).
93. B. Arkook, A. Alshahrie, N. Salah, M. Aslam, S. Aissan, A. Al-Ghamdi, A. Inoue, E.-S. Shalaan, Graphene & carbon nanotubes fibrous composite decorated with PdMg alloy nanoparticles with enhanced absorption–desorption kinetics for hydrogen storage application, *Nanomaterials*, 11, 2957 (2021).
94. Q.L. Lu, S.G. Huang, Y.D. Li, J.G. Wan, Q.Q. Luo, Alkali & alkaline-earth atom-decorated B38 fullerenes & their potential for hydrogen storage, *Int. J. Hydrogen Energy*, 40, 13022–13028 (2015).
95. Y. Zhang, X. Cheng, Hydrogen storage property of alkali & alkaline-earth metal atoms decorated C24 fullerene: A DFT study, *Chem. Phys.*, 505, 26–33 (2018).
96. R.K. Sahoo, B. Chakraborty, S. Sahu, Reversible hydrogen storage on alkali metal (Li & Na) decorated C20 fullerene: A density functional study, *Int. J. Hydrogen Energy*, 46, 40251–40261 (2021).
97. T. Wang, Z. Zhu, Stability & chaotic dynamic analysis of Li-doped fullerene-IRMOF composite materials for hydrogen storage, *Int. J. Hydrogen Energy*, 48, 11352–11369 (2023).

98. P. Liu, H. Zhang, X. Cheng, Y. Tang, Ti-decorated B38 fullerene: A high capacity hydrogen storage material, *Int. J. Hydrogen Energy*, 41, 19123–19128 (2016).

99. C. Tang, X. Zhang, The hydrogen storage capacity of Sc atoms decorated porous boron fullerene B40: A DFT study, *Int. J. Hydrogen Energy*, 41, 16992–16999 (2016).

100. P. Liu, H. Zhang, X. Cheng, Y. Tang, Yongjian Tang, Transition metal atom Fe, Co, Ni decorated B38 fullerene: Potential material for hydrogen storage, *Int. J. Hydrogen Energy*, 42, 15256–15261 (2017).

101. H.Y. Ammar, H.M. Badran, Ti deposited C20 & Si20 fullerenes for hydrogen storage application, DFT study, *Int. J. Hydrogen Energy*, 46, 14565–14580 (2021).

102. V. Mahamiya, A. Shukla, B. Chakraborty, Sc&ium decorated C24 fullerene as high capacity reversible hydrogen storage material: Insights from density functional theory simulations, *Appl. Surf. Sci.*, 573, 151389 (2022).

103. M. Shi, L. Bi, X. Huang, Z. Meng, Y. Wang, Z. Yang, Design of three-dimensional nanotube-fullerene-interconnected framework for hydrogen storage, *Appl. Surf. Sci.*, 534, 147606 (2020).

104. M. Gaboardi, N. Sarzi Amadé, M. Aramini, C. Milanese, G. Magnani, S. Sanna, M. Riccò, D. Pontiroli, Extending the hydrogen storage limit in fullerene, *Carbon*, 120, 77–82 (2017).

105. D.J. Durbin, N.L. Allan, C. Malardier-Jugroot, Molecular hydrogen storage in fullerenes - A dispersion-corrected density functional theory study, *Int. J. Hydrogen Energy*, 41, 13116–13130 (2016).

106. W. Huang, M. Shi, H. Song, Q. Wu, X. Huang, L. Bi, Z. Yang, Y. Wang, Hydrogen storage on chains-terminated fullerene C20 with density functional theory, *Chem. Phys. Lett.*, 758, 137940 (2020).

107. H. Ren, C. Cui, X. Li, Y. Liu, A DFT study of the hydrogen storage potentials & properties of Na- & Li-doped fullerenes, *Int. J. Hydrogen Energy*, 42, 312–321 (2017).

108. P. Liu, F. Liu, Q. Wang, Q. Ma, DFT simulation on hydrogen storage property over Sc decorated B38 fullerene, *Int. J. Hydrogen Energy*, 43, 19540–19546 (2018).

109. V. Mahamiya, A. Shukla, B. Chakraborty, Exploring yttrium doped C24 fullerene as a high-capacity reversible hydrogen storage material: DFT investigations, *J. Alloys Compd.*, 897, 162797 (2022).

110. J. Mao, P. Guo, T. Zhang, S. Zhang, C. Liu, A first-principle study on hydrogen storage of metal atoms (M = Li, Ca, Sc, & Ti) coated B40 fullerene composites, *Comp. Theor. Chem.*, 1181, 112823 (2020).

111. V. Tozzini, V. Pellegrini, Reversible hydrogen storage by controlled buckling of graphene layers, *J. Phys. Chem. C*, 115 25523 (2011).

112. Y. Miura, W. Dino, H. Nakanishi, First-principles studies for the dissociative adsorption of H2 on graphene, *J. Appl. Phys.*, 93, 3395 (2003).

113. O.K. Alekseeva, I.V. Pushkareva, A.S. Pushkarev, V.N. Fateev, Graphene and graphene-like materials for hydrogen energy. *Nanotechnol. Russia*, 15, 273–300 (2020).

114. Y.S. Al-Hamdani, A. Zen, A. Michaelides, D. Alfè, Mechanisms of adsorbing hydrogen gas on metal decorated graphene, *Phys. Rev. Mater.*, 7, 035402 (2023).

115. G.K. Sunnardianto, G. Bokas, A. Hussein, C. Walters, O.A. Moultos, P. Dey, Efficient hydrogen storage in defective graphene & its mechanical stability: A combined density functional theory & molecular dynamics simulation study, *Int. J. Hydrogen Energy*, 46, 5485–5494 (2021).

116. Z.M. Ao, F.M. Peeters, High-capacity hydrogen storage in Al-adsorbed graphene, *Phys. Rev. B*, 81, 205406 (2010).

117. W. Zhang, W.C. Lu, H.X. Zhang, K.M. Ho, C.Z. Wang, Hydrogen adatom interaction on graphene: A first-principles Study, *Carbon*, 131, 137–141 (2018).

118. V.V. Ivanovskaya, A. Zobelli, D. Teillet-Billy, N. Rougeau, V. Sidis, P.R. Briddon, Hydrogen adsorption on graphene: A first-principles study, *Eur. Phys. J. B*, 76, 481–486 (2010).

119. Y. Zhan, H. Cui, W. Tian, T. Liu, Y. Wang, Effect of hydrogen adsorption energy on the electronic & optical properties of Si-modified single-layer graphene with an Al decoration, *AIP Adv.*, 10, 045012 (2020).

120. E. Wu, C. Schneider, R. Walz, J. Park, Adsorption of hydrogen isotopes on graphene, *Nucl. Eng. Technol.*, 54, 4022–4029 (2022).

121. Y. Murata, A. Calzolari, S. Heun, Tuning hydrogen adsorption on graphene by gate voltage, *J. Phys. Chem. C*, 122, 11591–11597 (2018).

122. S. Rostami, A.N. Pour, M. Izadyar, Hydrogen adsorption by G-C3N4 & graphene oxide nanosheets, *J. Nanostruct.*, 9 (3), 498–509 (2019).

123. Y. Chen, Habibullah, G. Xia, C. Jin, Y. Wang, Y. Yan, Y. Chen, X. Gong, Y. Lai, C. Wu, Hydrogen storage properties of economical graphene materials modified by non-precious metal nickel & low-content palladium, *Inorganics*, 11, 251 (2023).

124. H. Wang, X. Yuan, Y. Wu, H. Huang, X. Peng, G. Zeng, H. Zhong, J. Liang, M. Ren, Graphene-based materials: Fabrication, characterization & application for the decontamination of wastewater & waste gas & hydrogen storage/generation, *Adv. Colloid Interface Sci.*, 195–196, 19–40 (2013).

125. H.G. Shiraz, O. Tavakoli, Investigation of graphene-based systems for hydrogen storage, *Renew. Sustain. Energy Rev.*, 74, 104–109 (2017).

126. K.K. Gangu, S. Maddila, S.B. Mukkamala, Characteristics of MOF, MWCNT & graphene containing materials for hydrogen storage: A review, *J. Energy Chem.*, 30, 132–144 (2019).

127. D. Feng, D. Zhou, Z. Zhao, T. Zhai, Z. Yuan, H. Sun, H. Ren, Y. Zhang, Progress of graphene & loaded transition metals on Mg-based hydrogen storage alloys, *Int. J. Hydrogen Energy*, 46, 33468–33485 (2021).

128. M. Singla, N. Jaggi, Theoretical investigations of hydrogen gas sensing & storage capacity of graphene-based materials: A review, *Sensors Actuators A: Phys.*, 332 (Part 1), 113118 (2021).

129. G. Sievi, D. Geburtig, T. Skeledzic, A. Bösmann, P. Preuster, O. Brummel, F. Waidhas, M.A. Montero, P. Khanipour, I. Katsounaros, J. Libuda, K.J.J. Mayrhofer, P. Wasserschei, Towards an efficient liquid organic hydrogen carrier fuel cell concept. *Energy Environ. Sci.*, 12 (7), 2305–2314 (2019).

130. D. Teichmann, K. Stark, K. Müller, G. Zöttl, P. Wasserscheid, W. Arlt, Energy storage in residential & commercial buildings via Liquid Organic Hydrogen Carriers (LOHC), *Energy Environ. Sci.*, 5, 9044–9054 (2012).

131. H. Imamura, Improved hydrogen storage by Mg & Mg–Nimodified by organic compounds, *J. Less Common Met.*, 172–174, 1064–1070 (1991).

132. H.G. Shiraz, M. Vagin, T.P. Ruoko, V. Gueskine, K. Karoń, M. Łapkowski, T. Abrahamsson, T. Ederth, M. Berggren, X. Crispin, Towards electrochemical hydrogen storage in liquid organic hydrogen carriers via proton-coupled electron transfers, *J. Energy Chem.*, 73, 292–300 (2022).

133. H. Yu, Y. Wu, S. Chen, Z. Xie, Y. Wu, N. Cheng, X. Yang, W. Lin, L. Xie, X. Li, J. Zheng, Pd-modified LaNi5 nanoparticles for efficient hydrogen storage in a carbazole type liquid organic hydrogen carrier, *Appl. Catal. B: Environ.*, 317, 121720 (2022).

134. M. Gambini, F. Guarnaccia, M. Manno, M. Vellini, Hydrogen flow rate control in a liquid organic hydrogen carrier batch reactor for hydrogen storage, *Int. J. Hydrogen Energy*, 51, 329–339 (2023).

135. S.P. Verevkin, A.A. Pimerzin, L.X. Sun, Liquid organic hydrogen carriers: Hydrogen storage by di-phenyl ether derivatives: An experimental & theoretical study, *J. Chem. Thermodynamics*, 144, 106057 (2020).

136. N. Heublein, M. Stelzner, T. Sattelmayer, Hydrogen storage using liquid organic carriers: Equilibrium simulation & dehydrogenation reactor design, *Int. J. Hydrogen Energy*, 45, 24902–24916 (2020).

137. J. Geiling, L. Wagner, F. Auer, F. Ortner, A. Nuß, R. Seyfried, F. Stammberger, M. Steinberger, A. Bösmann, R. Öchsner, P. Wasserscheid, K. Graichen, M. März, P. Preuster, Operational experience with a liquid organic hydrogen carrier (LOHC) system for bidirectional storage of electrical energy over 725 h, *J. Energy Storage*, 72 (Part D), 108478 (2023).

138. S.P. Verevkin, R.N. Nagrimanov, D.H. Zaitsau, M.E. Konnova, A.A. Pimerzin, Thermochemical properties of pyrazine derivatives as seminal liquid organic hydrogen carriers for hydrogen storage, *J. Chem. Thermodynamics*, 158, 106406 (2021).

139. M. Byun, A. Lee, S. Cheon, H. Kim, H. Lim, Preliminary feasibility study for hydrogen storage using several promising liquid organic hydrogen carriers: Technical, economic, & environmental perspectives, *Energy Convers. Manage.*, 268, 116001 (2022).

140. B.S. Shin, C.W. Yoon, S.K. Kwak, J.W. Kang, Thermodynamic assessment of carbazole-based organic polycyclic compounds for hydrogen storage applications via a computational approach, *Int. J. Hydrogen Energy*, 43, 12158–12167 (2018).

141. Y. Cao, H.A. Dhahad, S.G. Zare, N. Farouk, A.E. Anqi, A. Issakhov, A. Raise, Potential application of metal-organic frameworks (MOFs) for hydrogen storage: Simulation by artificial intelligent techniques, *Int. J. Hydrogen Energy*, 46, 36336–36347 (2021).

142. S.V. Vostrikov, M.E. Konnova, V.V. Turovtsev, S.P. Verevkin, Thermodynamics of hydrogen storage: LOHC system 1-alkyl-indole/octahydro-1-alkyl-indole, *Fuel*, 344, 128079 (2023).

143. S. Shirvani, D. Hartmann, K.J. Smith, Two-dimensional Mo2C: An efficient promoter for hydrogen storage & release from a liquid organic hydrogen carrier, *Int. J. Hydrogen Energy*, 48, 12309–12320 (2023).

144. Z. Ozturk, G. Ozkan, D.A. Kose, A. Asan, Experimental & simulation study on structural characterization & hydrogen storage of metal organic structured compounds, *Int. J. Hydrogen Energy*, 41, 8256–8263 (2016).

145. P. Makowski, A. Thomas, P. Kuhn, F. Goettmann, Organic materials for hydrogen storage applications: From physisorption on organic solids to chemisorption in organic molecules, *Energy Environ. Sci.*, 2, 480–490 (2009).

146. P.M. Modisha, C.N.M. Ouma, R. Garidzirai, P. Wasserscheid, D. Bessarabov, The prospect of hydrogen storage using liquid organic hydrogen carriers, *Energy Fuels*, 33 (4), 2778–2796 (2019).

147. Z. Abdin, C. Tang, Y. Liu, K. Catchpole, Large-scale stationary hydrogen storage via liquid organic hydrogen carriers, *iScience*, 24, 102966 (2021).

148. D. Zhao, X. Wang, L. Yue, Y. He, B. Chen, Porous metal–organic frameworks for hydrogen storage, *Chem. Commun.*, 58, 11059–11078 (2022).

149. D. Zhao, X. Li, K. Zhang, J. Guo, X. Huang, G. Wang, Recent advances in thermo-catalytic hydrogenation of unsaturated organic compounds with Metal-Organic Frameworks-based materials: Construction strategies & related mechanisms, *Coord. Chem. Rev.*, 487, 215159 (2023).

150. C. Chu, K. Wu, B. Luo, Q. Cao, H. Zhang, Hydrogen storage by liquid organic hydrogen carriers: Catalyst, renewable carrier, & technology - A review, *Carbon Resour. Convers.*, 6, 334–351 (2023).

151. K.C. Tan, Y.S. Chua, T. He, P. Chen, Strategies of thermodynamic alternation on organic hydrogen carriers for hydrogen storage application: A review, *Green Energy Resour.*, 1, 100020 (2023).

152. W. Bao, J. Yu, F. Chen, H. Du, W. Zhang, S. Yan, T. Lin, J. Li, X. Zhao, B. Zhu, Controllability construction & structural regulation of metal-organic frameworks for hydrogen storage at ambient condition: A review, *Int. J. Hydrogen Energy*, In Press (2023).

153. P. Muthukumar, A. Kumar, M. Afzal, S. Bhogilla, P. Sharma, A. Parida, S. Jana, E.A. Kumar, R.K. Pai, I.P. Jain, Review on large-scale hydrogen storage systems for better sustainability, *Int. J. Hydrogen Energy*, 48(15), 33223–33259 (2023).

154. Z. Xu, N. Zhao, S. Hillmansen, C. Roberts, Y. Yan, Techno-economic analysis of hydrogen storage technologies for railway engineering: A review, *Energies*, 15, 6467, (2022).
155. M. Sankir, N.D. Sankir (Edts.), *Hydrogen Storage Technologies*, Wiley & Sons, Inc. (2018).
156. T. Amirthan, M.S.A. Perera, The role of storage systems in hydrogen economy: A review, *J. Nat. Gas Sci. Eng.*, 108, 104843 (2022).
157. Y. Rong, S. Chen, C. Li, X. Chen, L. Xie, J. Chen, R. Long, Techno-economic analysis of hydrogen storage & transportation from hydrogen plant to terminal refueling station, *Int. J. Hydrogen Energy*, In Press (2023).

7 Hydrogen transport

7.1 INTRODUCTION

The green hydrogen has numerous methods of supply to distribute it to the end-users. These methods of supply make green hydrogen accessible for numerous applications in various industries. Hydrogen can be transported via dedicated pipelines, similar to how natural gas is dispersed. Hydrogen pipelines are utilized to move hydrogen from manufacturing facilities to various consumers, such as industrial facilities, hydrogen refueling stations, or other end-users. This method is effective for supplying hydrogen to sites with a consistent demand. The gaseous hydrogen can be compressed and stored in high-pressure tanks or cylinders for transport to end-users. This method is usually used to supply hydrogen to smaller facilities, laboratories, or for mobile applications like hydrogen fuel cell vehicles. Hydrogen can be cooled and liquefied at very low temperatures (around $-253°C$ or $-423°F$) to become liquid hydrogen (LH_2). LH_2 is more energy-dense than gaseous hydrogen, making it appropriate for long-distance transport. It is characteristically transported in specialized cryogenic tanker trucks or containers [1].

For hydrogen fuel cell vehicles, dedicated hydrogen refueling stations are utilized to supply gaseous hydrogen. These stations produce hydrogen on-site through electrolysis or receive hydrogen deliveries in gaseous or liquid form and dispense it to vehicles as required. In the maritime sector, hydrogen bunkering is a method used to supply hydrogen to ships. This procedure comprises refueling ships with gaseous or LH_2 at specialized bunkering facilities situated at ports. These systems comprise transporting compressed hydrogen gas in a tube trailer, which is fundamentally a high-pressure tube or container. The tube trailer can be transported to numerous places, making it useful for supplying hydrogen to zones without dedicated pipelines. The choice of the method of supply depends on numerous factors, including the volume and distance of hydrogen transport, the infrastructure existing, the precise needs of end-users, and the economic feasibility of the chosen method. As green hydrogen production and demand grow, there will likely be development and innovation in these manners of supply to ensure the effective and widespread distribution of this clean energy source [2].

7.2 PIPELINE TRANSPORT

The pipeline transportation of green hydrogen comprises the use of dedicated pipelines to transport hydrogen gas from manufacturing facilities to end-users or storage facilities [3]. As per the European Commission Joint Research Centre, the best option for transporting hydrogen is the hydrogen pipeline for distances beneath 6,500 km. The shipping of liquefied hydrogen for distances between 6,500 km and 10,000 km is also feasible. The liquid organic hydrogen carrier for distances above

DOI: 10.1201/9781003432364-9

10,000 km is used for transportation. The dedicated pipelines are used to transport the manufactured hydrogen to its destination. These pipelines are naturally made of materials compatible with hydrogen, such as steel or composite materials. The hydrogen pipelines need to meet safety and material compatibility standards to stop leaks and confirm safe transportation. Pipeline transport is most effective for long distances, which makes it appropriate for transporting green hydrogen from remote renewable energy facilities to zones with high hydrogen demand. The efficacy of hydrogen pipelines is affected by factors like distance, pressure, and pipeline diameter. Hydrogen may require to be compressed before injection into the pipeline and decompressed at the receiving end to meet the mandatory pressure levels. The compression and decompression processes consume energy and must be wisely managed to minimize energy losses.

European Commission Joint Research Centre published a report in 2021 describing different hydrogen transport options from an economic point of view. The report conducted a survey on hydrogen transport among the different parts of Southern and Central Europe with 2,500 km for transporting hydrogen for 270 hydrogen refueling stations located in a 500 km radius [4]. Later on, in 2022, the European Commission Joint Research Centre published a detailed report with a main emphasis on distance analysis for 1 Mt/y. Also, the report showed hydrogen pipeline sensitivity analysis to various targeted amounts of yearly delivered hydrogen for 2,500 km [5] that helped to describe some of the effects of transport rate.

More than 2,575 km pipelines are available in the world for hydrogen distribution, mainly located near the chief hydrogen customers like chemical plants, oil refineries, etc. [6]. The ductile steel pipelines present embrittlement problems at high pressures and mass flow rates [7]. It is supposed that for pure hydrogen, a polymer like polyethylene may be a good alternative. This is due to the extrusion feature which can decrease engineering price. Besides, this material can be used easily with many required joints [8]. Nevertheless, sometimes, the hydrogen leakage problem exists [9,10]. This problem may be overcome by making a composite material of polyethylene, providing high strength to the pipeline and reducing buoyancy in deep waters [11]. Normally, ground-based pipelines are used at 10–30 bar and, nowadays, scientists are advising using this ground-based pipeline at 60–100 bar pressure with 10–15 m/s flow rates [12]. The submarine hydrogen pipelines could not attain commercial applications due to certain problems [13]. However, the DNV standard for submarine pipeline systems (DNV-ST-F101) is a well-known standard for reliable design and safety. But still, we need to make sure of the safety, leakage, and economic issues of DNV-ST-F101.

Safety is a serious concern in the transport of hydrogen. Hydrogen is highly combustible, and pipeline systems must include safety measures like leak detection, venting systems, and robust materials to mitigate risks. The advanced control and checking systems are important for maintaining the reliability and security of hydrogen pipelines. These systems aid operators in detecting leaks, controlling pressure, and confirming the pipeline's efficient operation. As the demand for green hydrogen upsurges, the pipeline infrastructure may need to be extended and optimized. Investments in pipeline networks are needed to support the development of the hydrogen economy. The hydrogen pipelines can be combined with other energy infrastructure, like natural gas pipelines. This can ease the blending of hydrogen into existing gas grids, donating a way to lead hydrogen into the energy mix. The transport of green hydrogen may be subject to guidelines and standards set by local, national, and international authorities. Compliance with safety standards as well as environmental and safety assessments is essential. The pipeline transport of green hydrogen has the potential to play an important role in decarbonizing numerous sectors, including transportation, industry, and power generation by providing a clean and sustainable energy carrier for these applications. However, the development of such infrastructure requires careful planning and investment to ensure safety and efficiency.

7.3 CRYOGENIC LIQUID TANKER TRUCKS

Trucks transporting LH_2 are denoted as liquid tankers [14]. The cryogenic liquid tanker trucks are particular vehicles intended to transport and supply cryogenic liquids, which are substances that exist at extremely low temperatures, typically below $-150°C$ ($-238°F$). These trucks are important for the safe and effective distribution of cryogenic gases and liquids such as LH_2, liquid nitrogen, liquid oxygen, and liquid natural gas (LNG). The cryogenic tanker trucks are equipped with specific, double-walled tanks that are heavily insulated to keep the cryogenic liquids at very low temperatures. The inner tank is naturally made of aluminum or stainless steel, and the outer tank is utilized as a secondary barrier to lessen heat transfer. These liquid tanker trucks are classically loaded at dedicated cryogenic production facilities or storage tanks, which are frequently situated at industrial gas plants. The loading procedure comprises transferring the cryogenic liquid from the storage tank to the tanker's insulated tank, and this is classically done via a pumping system. Safety is of supreme rank when dealing with cryogenic liquids owing to their great cold and potential dangers. These tanker trucks are equipped with safety structures such as pressure relief valves, emergency venting systems, vacuum insulation, and alarms to monitor pressure and temperature.

The cryogenic liquid tanker trucks are precisely intended to minimize heat transfer and the boil-off of cryogenic liquids in transport. These frequently have a cryogenic refrigeration system to keep low temperatures and decrease losses. The trucks are also prepared with advanced control systems to monitor and continue the desired pressure and temperature conditions inside the tank. When the cryogenic tanker truck reaches its target, it can either directly supply cryogenic liquids to end-users or transfer them to on-site storage tanks. The cryogenic liquids are naturally unloaded using a pump or a gravity-fed system. Cryogenic liquid transport is subject to severe safety regulations and standards to ensure the protection of both the public and the environment. Compliance with regulations is important [15].

The chart industry specializes in manufacturing cryogenic equipment that provides equipment solutions for the full supply chain of hydrogen fuel, comprising onboard vehicle tanks, transport equipment, fuel stations, and LH_2 production. These industries have been designing and building LH_2 trailers for several years. Also, SAG is a technology leader in the field of cryogenic tanks for LH_2 and LNG. It has developed the first LH_2 cryogenic tank system for trucks. Air Liquide and Faurecia are bringing LH_2 to heavy-duty vehicles with 1,000 km range. This also includes the optimized storage capacity and tested technology ready within five years. Cryogenic liquid tanker trucks play an important role in supplying cryogenic gases and liquids for a wide range of industrial and commercial applications. These vehicles permit the safe and effective distribution of cryogenic substances that are important for numerous sectors, comprising healthcare, manufacturing, and energy production.

Generally, the tank sizes can vary from 1.5 m³ (100 kg) to 75.0 m³ (5,000 kg). The single largest LH_2 tank was used at the John F. Kennedy Space Center, USA. This comprises 900,000 gallons of LH_2. This tank has been utilized for Apollo rockets fueling [16]. The cryogenics procedure keeps the hydrogen in liquid form by cooling it to an extreme temperature, i.e., below −253°C (−423°F). The Kawasaki Heavy Industries (KHI) in Japan is encouraging a project for hydrogen liquefaction. The hydrogen being used is prepared by gasification/CCS of brown coal in Australia following the transportation to Japan by LH_2 tankers. So, this may help to create a hydrogen supply chain that is needed for hydrogen society development in Japan. The researchers at these industries had started pilot operations in 2017 utilizing a 2,500 m³ prototype tanker. This is providing marketable service and is expected to reach its goals by 2025 using 160,000 m³ large tankers [17].

7.4 HYDROGEN SHIPPING

In 2012, Rohde and Sames [18] developed a model design for a zero-emission vessel on a ship. The ship was 137.22 m in length with 1,000 TEU (twenty-foot equivalent unit) overall hydrogen intake. A hybrid scheme made of fuel cells and batteries was used to produce electricity for both onboard energy supply and propulsion. WE-NET program developed a theoretical design for a 200,000 m³ hydrogen tanker. This was based on LNG ship knowledge. Some other studies on hydrogen tank insulation were also carried out [19]. Alkhaledi et al. [20] discussed a detailed theoretical evaluation and view of a feasible design for a large LH_2 tanker fueled by LH_2. The size of the tank, stability of the ship, and characteristics of the ship were utilized to assess the initial design and performance of the liquefied hydrogen tanker. This tanker was given the name "JAMILA." This was prepared by four big LH_2 tanks with a total capacity of approximately 280,000 m³. The tanker was loaded on a ship 370 m long and 75 m wide with 20,000 tons of hydrogen-carrying capacity. In 2021, KHI obtained the Approval in Principle for an LH_2 ship design with a size of 160,000 m³ from the Classification Society Class NK that efficiently utilizes the boil-off gas to run the ship. In 2022, the Suiso Frontier, prepared by KHI, delivered its first LH_2 shipment to Kobe in Japan after filling from Hastings in Australia.

7.5 GASEOUS TUBE TRAILERS

The green hydrogen supply by gaseous tube trailers denotes a method of transporting and distributing green hydrogen gas using particular trailers furnished with high-pressure gas cylinders or tubes. The gaseous tube trailers are a good way to transport and distribute the green hydrogen to numerous locations. The high-purity hydrogen gas is compacted and filled into high-pressure gas cylinders or tubes,

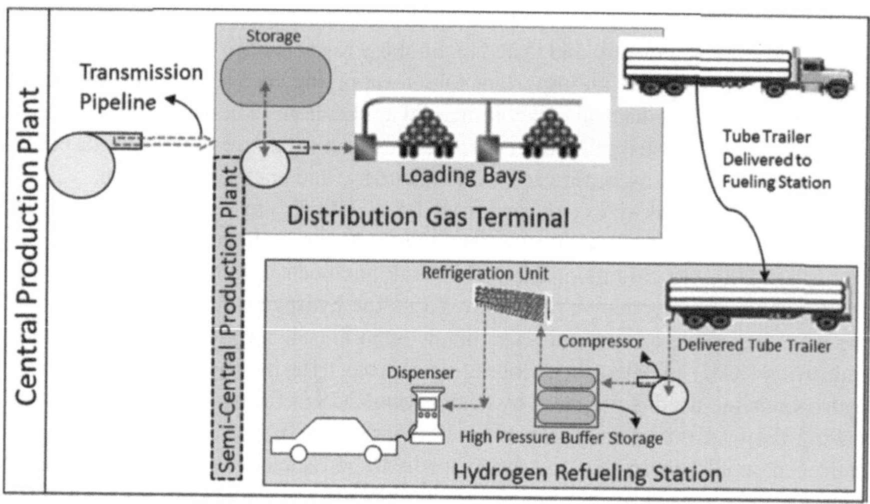

FIGURE 7.1 A schematic representation of the hydrogen tube trailer delivery pathway [reproduced with permission from ref. 24].

classically made of composite materials. These filled tubes or cylinders are loaded onto specialized trailers, which are planned to transport them to the end-users or distribution points. These trailers frequently comprise safety features to stop leaks and confirm secure transportation. The gaseous tube trailers are then transported to numerous places, such as refueling stations, industrial facilities, and other points of use. Upon arrival, hydrogen can be offloaded and distributed to clients, where it can be utilized for numerous applications, such as fuel cell vehicles, industrial processes, power generation, etc. [21–23].

The green hydrogen can be transported using tube trailers, i.e., trucks hauling gaseous hydrogen compressed to 180 bar (~2,600 psi) pressures or higher into long cylinders. These are stacked on a trailer that the truck hauls. Currently, the tube trailers are restricted to pressures of 250 bar by the U.S. Department of Transportation regulations, but exceptions have been granted to enable operation at higher pressures (e.g., 500 bar or higher). The steel tube trailers are most regularly engaged and carry approximately 380 kg on board, while composite storage containers have been developed with capacities of 560–900 kg of hydrogen per trailer. The use of tube trailers may be the first step in constructing the hydrogen infrastructure. These may be moved to serve diverse locations as required. Tube trailers can transport green hydrogen from areas of highly renewable resources, including remote areas with little or no hydrogen demand, to demand centers, making it a truly global energy solution [24,25]. A project in Queensland (Australia) is working on producing green hydrogen. Green hydrogen is being supplied in remote areas via tube trailers. The action of high-pressure tubes or new techniques in those areas may cause some problems. That is why 300 bar pressure tube trailers have been selected for this purpose [23]. The schematic representation of the hydrogen tube trailer delivery pathway is shown in Figure 7.1.

The hydrogen transport by tube trailers can be compressed further and stored in a high-pressure buffer at the refueling station. The use of gaseous tube trailers for green hydrogen transportation provides numerous advantages. The usage of high-pressure storage containers on tube trailers confirms the safe transport of hydrogen gas. The tube trailers deliver flexibility in distributing hydrogen to remote or temporary sites, making it a feasible option for areas without devoted hydrogen pipelines. Hydrogen is manufactured using renewable energy sources, reducing greenhouse gas emissions. This can play a role in decarbonizing industries and transportation. However, there are also challenges, such as transportation costs, the energy required for compression, and the need for an established infrastructure for filling and refilling tube trailers [26]. Briefly, green hydrogen supply by gaseous tube trailers is one of the methods to transport and deliver ecologically friendly hydrogen from manufacturing sites to end-users or circulation points, supporting the growth of the green hydrogen industry and contributing to sustainable energy and environmental goals.

REFERENCES

1. A.H. Azadnia, C. McDaid, A.M. Andwari, S.E. Hosseini, Green hydrogen supply chain risk analysis: A European hard-to-abate sectors perspective, *Renew. Sustain. Energy Rev.*, 182, 113371 (2023).
2. W. Dong, C. Shao, X. Li, D. Zhu, Q. Zhou, X. Wang, Integrated planning method of green hydrogen supply chain for hydrogen fuel cell vehicles, *Int. J. Hydrogen Energy*, 48, 18385–18397 (2023).
3. R. D'amore-Domenech, V.L. Meca, B.G. Pollet, T.J. Leo, On the bulk transport of green hydrogen at sea: Comparison between submarine pipeline and compressed and liquefied transport by ship, *Energy*, 267, 126621 (2023).
4. C.R. Ortiz, F. Dolci, E.W. Ronnefeld, *Assessment of Hydrogen Delivery Options, EUR 31199 EN*, Publications office of the European Union (2022).
5. R.O. Cebolla, F. Dolci, E. Weidner, *Assessment of Hydrogen Delivery Options: Feasibility of Transport of Green Hydrogen Within Europe*, Publications Office of the European Union Commission (2022).
6. O.S. Ibrahim, A. Singlitico, R. Proskovics, S. McDonagh, C. Desmond, J.D. Murphy, Dedicated large-scale floating offshore wind to hydrogen: Assessing design variables in proposed typologies. *Renew. Sustain. Energy Rev.*, 160, 112310 (May 2022).
7. O. Bouledroua, Z. Hafsi, M.B. Djukic, S. Elaoud, The synergistic effects of hydrogen embrittlement and transient gas flow conditions on integrity assessment of a precracked steel pipeline. *Int. J. Hydrogen Energy*, 45(35), 18010–18012 (2020)
8. Y. Su, H. Lv, W. Zhou, C. Zhang, Review of the hydrogen permeability of the liner material of type iv on-board hydrogen storage tank. *World Electr. Veh. J.*,12 (3), 1–18 (2021).
9. Z.L. Messaoudani, F. Rigas, M.D.B. Hamid, C.R.C. Hassan, Hazards, safety and knowledge gaps on hydrogen transmission via natural gas grid: A critical review. *Int. J. Hydrogen Energy*, 41(39), 17511–17525 (2016).
10. M-H. Klopffer, P. Berne, É. Espuche, Development of innovating materials for distributing mixtures of hydrogen and natural gas. Study of the barrier properties and durability of polymer pipes, *Oil Gas Sci. Technol. Rev. d'IFP Energies Nouv*, 70(2) , 305–315 (2015).
11. M.J. Kaiser, Offshore pipeline construction cost in the U.S. Gulf of *Mexico. Mar. Policy*, 82, 147–166 (August 2017).
12. F.H. Saadi, N.S. Lewis, E.W. McFarland. Relative costs of transporting electrical and chemical energy. *Energy Environ. Sci.* 11 (3), 469–475 (2018).

13. O.S. Ibrahim, A. Singlitico, R. Proskovics, S. McDonagh, C. Desmond, J.D. Murphy, Dedicated large-scale floating offshore wind to hydrogen: Assessing design variables in proposed typologies. *Renew. Sustain. Energy Rev.*, 160, 112310 (2022).
14. B. Zohuri, *Physics of Cryogenics, An Ultralow Temperature Phenomenon*, 1st Edn., Elsevier (2017).
15. Y. Yan, Z. Xu, F. Han, Z. Wang, Z. Ni, Energy control of providing cryo-compressed hydrogen for the heavy-duty trucks driving, *Energy*, 242, 122817 (2022).
16. B. Zohuri, *Hydrogen Energy: Challenges and Solutions for a Cleaner Future*, Springer (2019).
17. K. Sasaki, H.-W. Li, A. Hayashi, J. Yamabe, T. Ogura, Stephen M. Lyth (Edts.), *Hydrogen Energy Engineering: A Japanese Perspective*, Springer (2016).
18. F. Rohde, P. Sames, Conceptual design of a zero-emission open-top container feeder. In *Proceedings of the 8th International Conference on High-Performance Marine Vehicles (HIPER)*, Duisburg, Germany, 27–28 September 2012; 207–215 (2012).
19. A. Abe, M. Nakamura, I. Sato, H. Uetani, T. Fujitani, Studies of the large-scale sea transportation of liquid hydrogen, *Int. J. Hydrogen Energy*, 23, 115–121(1998).
20. A.N. Alkhaledi, S. Sampath, P. Pilidis, A hydrogen fuelled LH_2 tanker ship design. *Ships Offshore Struct.*, 17, 1555–1564 (2021).
21. A. Elgowainy, K. Reddi, E. Sutherland, F. Joseck, Tube-trailer consolidation strategy for reducing hydrogen refueling station costs. *Int. J. Hydrogen Energy*, 39, 20197–20206 (2014).
22. M. Yang, R. Hunger, S. Berrettoni, B. Sprecher, B, Wang, A review of hydrogen storage and transport technologies, *Clean Energy*, 7, 190–216 (2023).
23. T. Hasan, N.M.S. Hassan, R. Shah, K. Emami, J. Anderson, A hydrogen supply-chain model powering Australian isolated communities, *Energy Rep.*, 9, 209–214 (2023).
24. K. Reddi, A. Elgowainy, N. Rustagi, E. Gupta, Techno-economic analysis of conventional and advanced high-pressure tube trailer configurations for compressed hydrogen gas transportation and refuelling, *Int. J. Hydrogen Energy*, 43, 4428–4438 (2018).
25. M.G. Güler, E. Geçici, A. Erdoˇgan, Design of a future hydrogen supply chain: A multi period model for turkey. *Int. J. Hydrogen Energy*, 46, 16279–16298 (2021).
26. V. Vijayakumar, A. Jenn, J. Ogden, Modeling future hydrogen supply chains in the western United States under uncertainties: An optimization-based approach focusing on California as a hydrogen hub, *Sustain. Energy Fuels*, 7, 1223 (2023).

8 Market and impact on society

8.1 INTRODUCTION

The market for green or clean hydrogen, frequently referred to as "green hydrogen," has been attaining noteworthy attention and momentum in recent years as countries and industries aim to decrease carbon emissions and transition to more sustainable energy sources. The most significant aspects of the green energy hydrogen market are fast growth, end-user sectors, industries, power cohort, energy storage, manufacturing technologies, cost decrease, investment and collaboration, infrastructure development, global players, environmental benefits, competitive landscape, and policy and guidelines.

The green hydrogen market is facing fast growth owing to increased environmental concerns and the necessity to decarbonize numerous sectors. Governments, businesses, and investors are paying strong attention to green hydrogen projects and techniques. Green hydrogen is utilized in fuel cell electric vehicles (FCEVs) and can play a critical role in decarbonizing the transportation sector. Green hydrogen can be utilized in industries such as chemicals, steel manufacturing, and petrochemicals, helping to decrease carbon emissions in these energy-intensive segments. It can be used in fuel cells for power cohort, especially during peak mandate and to balance intermittent renewable energy sources. Green hydrogen can aid as a means of energy storage for additional renewable energy, which can be used during periods of high mandate.

One of the challenges of green hydrogen has been its comparatively high manufacturing cost compared to other forms of hydrogen. However, ongoing research, technical progress, and economies of scale are driving down the cost of green hydrogen manufacture, making it more inexpensive. There is a noteworthy investment in green hydrogen projects, including partnerships between governments, energy companies, and research organizations. The collaboration at a global scale is important to advance the green hydrogen industry. To fully recognize the potential of green hydrogen, widespread infrastructure development is mandatory, including manufacturing facilities, storage, transportation, and distribution networks. These progress are being planned and applied in many regions.

Green hydrogen is assumed an environmentally friendly energy carrier as its production creates no greenhouse gas releases. It can play an important role in achieving carbon neutrality goals. Green hydrogen is recognized as competition from other forms of clean energy, such as battery technology and carbon capture and application. The industry's future success depends on its capability to remain competitive with these substitutes. Several countries, including Australia, Germany, Japan,

DOI: 10.1201/9781003432364-10

and the European Union, have outlined ambitious green hydrogen plans. Major energy companies and manufacturers are also investing in green hydrogen projects. Governments worldwide are applying policies and regulations to support the green hydrogen marketplace. The subsidies, incentives, and targets for green hydrogen manufacture and utilization are becoming more common to boost its adoption. The green hydrogen market is poised to become a critical component of the clean energy transition, as it can address the necessity for clean and reliable energy across numerous sectors. As technology progresses and manufacturing costs continue to decline, green hydrogen is expected to play a vibrant role in dipping carbon emissions and attaining a more sustainable energy future. The worldwide green hydrogen market was estimated at USD 242.7 billion in 2023 and is predicted to reach USD 410.6 billion by 2030, rising at 7.8% compound annual growth rate (CAGR) during the prediction period [1].

8.2   INDUSTRIES

The industrial market for green hydrogen states the utilization of hydrogen manufactured using renewable energy sources in numerous industrial applications. Numerous industries use hydrogen for several purposes, including petroleum refining, ammonia production, chemicals manufacturing, and shipping industries [2]. In these sectors, hydrogen is usually manufactured from natural gas, resulting in carbon emissions. The changeover to green hydrogen can considerably reduce the carbon footprint of these industries.

The steel industry is a main customer of hydrogen. The industry can considerably cut its CO_2 emissions and changeover to a more sustainable and environmentally friendly operation by using green hydrogen in steel production processes. Green hydrogen can be utilized in the manufacture of ammonia, which is vital for fertilizers. This application has the talent to decrease the carbon footprint of agriculture, donating to sustainable food manufacturing. Hydrogen is used as a feedstock in the chemical and petrochemical industries. Shifting to green hydrogen in these procedures can lessen emissions and encourage the expansion of sustainable and low-carbon chemical goods. The use of green hydrogen in industry considerably decreases greenhouse gas emissions, which aligns with global environmental aims and aids companies in meeting regulatory necessities related to emissions reduction. Green hydrogen can increase the energy efficacy of industrial processes by serving as a clean energy source. It can also be utilized for power generation in fuel cells, serving industries to balance grid mandates and integrate cleaner energy sources.

The progress of the green hydrogen industry can lead to financial development, job creation, and investment in research and development in the industrial sector. Green hydrogen's widespread acceptance in industrial applications frequently needs international collaborations and corporations among companies, countries, and research institutions. The growth of the market of hydrogen can be augmented by sharing knowledge and resources. The establishment of a green hydrogen infrastructure is critical for its placement in industrial applications. This includes the growth of production facilities, transportation systems, and storage solutions. Governments are introducing incentives, policies, and regulations to inspire the use of green hydrogen in industrial procedures. This includes tax incentives, financial support, and emission decrease targets.

The continuous research and development efforts are focused on refining the efficacy, scalability, and cost-effectiveness of green hydrogen manufacture and use in industry. Companies in the industrial sector are identifying the profits of transitioning to green hydrogen. Those who invest in these sustainable tasks can gain a competitive benefit in the market by dipping their carbon footprint and meeting client demand for environmentally accountable products. The industrial market for green hydrogen is a critical component of the broader effort to decarbonize industrial methods and decrease emissions. The industries can take an important step toward sustainability by adopting green hydrogen, while also contributing to worldwide efforts to combat climate alteration.

Briefly, the market was valued at USD 3.2 billion in 2021 and is predicted to reach USD 7,314 million by 2027, growing at a CAGR of 61.0%. The mobility industry accounted for a share of 58% in terms of value in the Green Hydrogen Market in 2022 and is predicted to reach USD 4,550 million by 2027 at a CAGR of 63.4%. The power sector is expected to the second-highest CAGR of 63.0%, rising from USD 88.5 million in 2022 to USD 1,018 million by 2027 [3].

8.3 TRANSPORTATION

The worldwide hydrogen storage tanks and transport market was assessed at USD 174 million in 2022 and is predicted to go to USD 4,155 million by 2030, rising at a CAGR of 48.6%. The transport market for green hydrogen refers to the utilization of hydrogen manufactured by renewable energy sources. Green hydrogen is seen as a talented solution to decarbonize the transport sector, which is a main contributor to greenhouse gas releases [4,5]. The most significant features of the green energy hydrogen market in transportation are zero emissions, heavy-duty transportation, FCEVs, maritime and aviation, public transportation, energy efficiency, hydrogen refueling infrastructure, decreased noise contamination, economic progress and job creation, and global partnership.

Green hydrogen has applications in heavy-duty transport, such as trucks, buses, and even trains. It can offer an effective and clean energy source for long-haul and high-load vehicles. Green hydrogen can be utilized in FCEVs. FCEVs have the benefit of long driving ranges and quick refueling times, making them a possible alternative to classical inner combustion engine vehicles. FCEVs are known for their energy competence. They change hydrogen into electricity via fuel cells, which makes them very effective and can help decrease overall energy intake in the transport sector. FCEVs powered by green hydrogen yield zero tailpipe emissions. FCEVs are lower than traditional internal combustion engine vehicles, decreasing noise pollution in urban areas. This can significantly decrease air pollution in urban areas and help diminish climate change by reducing CO_2 releases. The basic structure of an FC-based electrical vehicle is shown in Figure 8.1 [6] while Figure 8.2 shows an FCEV structure diagram [7].

Green hydrogen can be utilized in maritime and aviation industries to decrease releases in shipping and air travel. Hydrogen-powered ships and airplanes are in

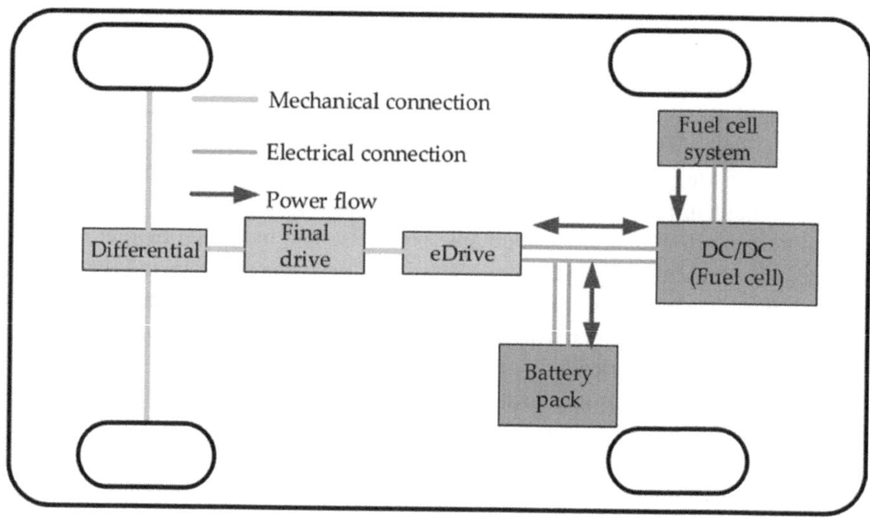

FIGURE 8.1   Basic fuel cell hybrid electric vehicle transmission structure diagram [reproduced with permission from ref. 6].

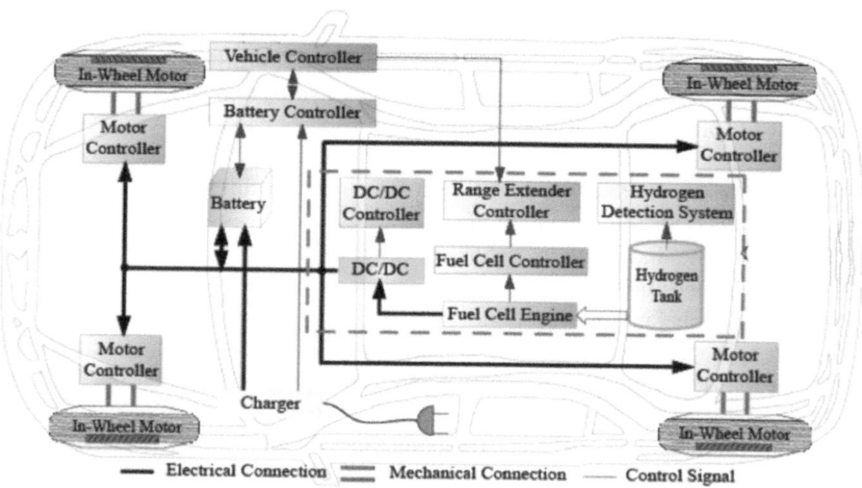

FIGURE 8.2   Extended-range fuel cell electric vehicle structure diagram [reproduced with permission from ref. 7].

progress and have the future to lower the carbon footprint of these modes of transport. The public transport systems in cities are progressively adopting hydrogen fuel cell buses and other vehicles to shrink releases and improve air quality. The growth of green hydrogen in transport relies on the development of refueling infrastructure, including hydrogen stations and transport chains. These infrastructures are critical

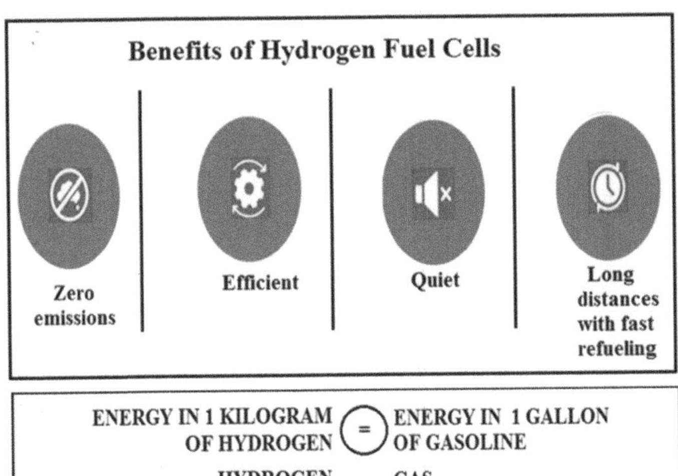

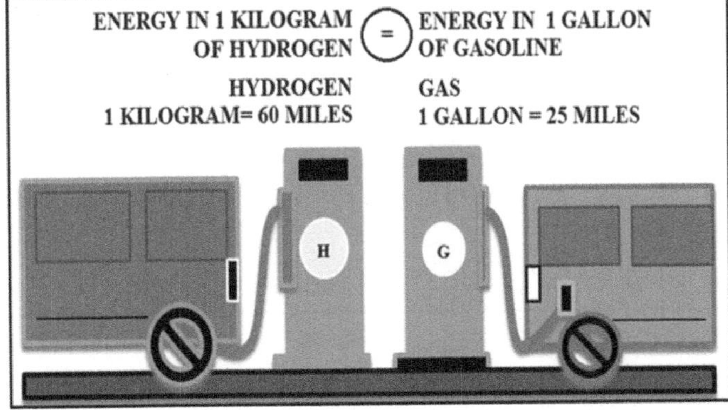

FIGURE 8.3 Advantages of green energy hydrogen in transportation.

for the widespread adoption of FCEVs. Some of the advantages of green energy hydrogen in transportation are shown in Figure 8.3.

Many nations are investing in green hydrogen research and set-up development for transport, fostering international collaboration and partnerships to quicken adoption. Governments are introducing incentives and policies to support green hydrogen adoption in the transport sector, including tax incentives, subsidies, and release saving targets. The utilization of green hydrogen in transport aligns with countries' sustainability and climate aims, helping them decrease their carbon footprint and meet international climate contracts. In spite of many benefits green hydrogen has competition from other zero-emission technologies like battery electric vehicles. The growth of green hydrogen in transport is also reliant on decreasing manufacturing costs and refining infrastructure. The transport market for green hydrogen denotes a noteworthy opportunity to decrease releases and improve the sustainability of the transport sector. As technology progresses, cost reduces, and infrastructure enlarges, green hydrogen is poised to perform an important role in the changeover to cleaner and more sustainable transportation options. The hydrogen refueling stations at the end of 2018 are shown in Figure 8.4 [8]. This clearly indicates the growth and need for green energy hydrogen. The road vehicle fleet predicted by 2030 is shown in

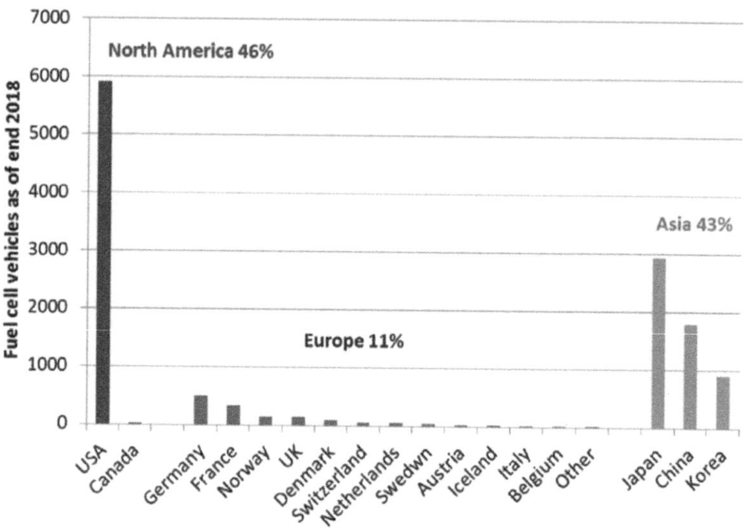

FIGURE 8.4   Hydrogen refueling stations at the end of 2018 [reproduce with permission from ref. 8].

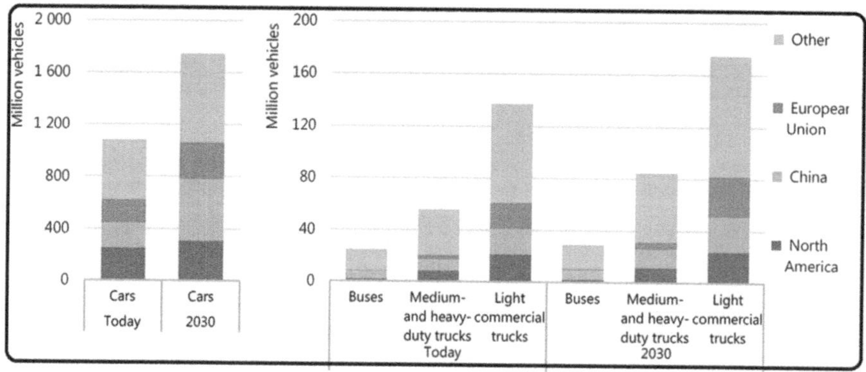

FIGURE 8.5   The road vehicle fleet prediction by 2030 [reproduced with permission from ref. 9].

Figure 8.5. These data are a clear indication of the growth of the transportation in near future.

The designs of numerous cars such as the Toyota Mira I, Honda FCX Clarity, the Honda Clarity FCEV, the Hyundai Nexo, the Mercedes Benz B-class FCEV, the Chevrolet Equinox Fuel Cell, the Toyota Mirai II, and the Hyundai ix35 (an improved version of the Hyundai Tucson) are shown in Figure 8.6. All shown vehicles are assumed to be midsized cars with almost like and power outputs. It should be observed that none of the cars shown are PHEVs since their battery capabilities

FIGURE 8.6 Midsized cars with less than 2kW battery capability: (a) Honda FCX clarity, (b) Hyundai Tucson/ix35 FC, (c) Toyota Mirai I, (d) Honda clarity FC, (e) Hyundai Nexo, (f) Mercedes Benz F-Cell, (g) Chevrolet Equinox Fuel Cell, and (h) Toyota Mirai II [10].

are less than 2 kWh [10]. A comparison of the key factors of the hydrogen fuel cell-powered cars, which are commercially accessible on the market, is shown in Table 8.1. This table clearly indicates the different ranges, power output, hydrogen tank weight, and price for different models and manufacturing years.

8.4 SPACE EXPLORATION

The space investigation market is developing and exciting, especially with green hydrogen [11]. The use of green hydrogen in space research can bring numerous benefits and revolutions. The most significant aspects of the green energy hydrogen market in space exploration are reusable rockets, deep space exploration, propulsion systems, propulsion systems, resource utilization on other planets, international collaboration, in-situ resource utilization, long-duration missions, reduced cost and environmental impact, research and development and planetary colonization, and reduced emissions.

TABLE 8.1

A comparison of some features of some automobiles

Production year	Vehicle model	Range (km/miles)	Power output (kW/hp)	Hydrogen tank weight (kg)/ capacity (L)	Price (USD)
2008	Honda FCX Clarity	450/270	100/134	4.1/171	34.995
2013	Hyundai Tucson /ix 35 Fuel Cell	415/258	124/-	5.6/140	-
2014	Toyota Mirai	502/312	114/153	5/122	57.500
2016	Honda Clarity Fuel Cell-FCEV	740/460	105/-	NA	- -
2018	Hyundai Nexo	609/378	135/-	6.3/156	59.435
2010	Mercedes Benz B-Class F-Cell	400/250	100/136	3.7/-	-
2007	Chevrolet Equinox Fuel Cell	320/200	94/126	4.2/-	-
2020	Toyota Mirai II	650/404	128/182	NA	49.500

Green hydrogen can be utilized as a clean and effective fuel for rocket propulsion. Hydrogen-powered rocket engines can deliver high thrust and are appropriate for numerous space missions, including crewed missions to the Moon and Mars. Green hydrogen technology could be utilized to extract hydrogen from celestial sources like lunar ice or Martian water for utilization in propulsion and power generation, dipping the need to transport fuel from Earth. Green hydrogen manufacture on other celestial bodies using in-situ resource use can enable self-sustaining missions and offer resources for manifold purposes, including power, propulsion, and life support. The reusability of rockets is essential for cost-effective space research. Green hydrogen-powered rockets have the capacity to be reused manifold times, dipping launch costs and making space exploration more economically feasible. Hydrogen fuel cells can be utilized to power spacecraft and habitats for deep space research missions, offering a consistent and sustainable source of energy in the tough environment of space. The utilization of green hydrogen in rocket propulsion results in water vapor as the main byproduct. This decreases the environmental effect of space launches, particularly in terms of emissions into Earth's atmosphere.

The European Space Agency is working on a project called HYdrogène GUyanais A Neutralité Environnementale to create a pilot plant within the Spaceport capable of producing 130 tons per year of green hydrogen. The goal of this project is to reduce spaceport greenhouse gas emissions. The project aims to replace "gray" hydrogen, (manufactured by a CO_2-intensive industrial process known as methanol steam reforming) with green hydrogen (manufactured by water electrolysis from renewable energy). The green hydrogen will be utilized to fuel rockets, dipping CO_2 releases linked to the Ariane program by several thousand tons per year. An Austrian start-up (HydroSolid GmbH) is an alternative to develop green hydrogen uses in the space. The utilization of liquid hydrogen in space is shown in Figure 8.7.

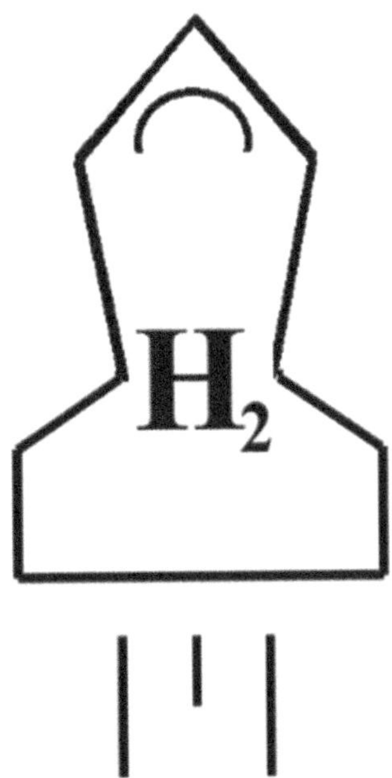

FIGURE 8.7 Liquid hydrogen use in space.

Hydrogen fuel cells can offer a long-lasting and effective source of power for missions with extended periods, such as those concerning rovers and habitats on other heavenly bodies. The space research market frequently includes international partnerships, and green hydrogen technology can be a point of collaboration in global space activities. The sustainability and compact environmental influence of green hydrogen technology can support space agencies' objectives to curtail the environmental special effects of space investigation activities. The space investigation market can drive research and development efforts in green hydrogen technology, leading to progress that can be advantageous to Earth-based applications. Green hydrogen technology can play a role in supporting future colonization settlement efforts by providing a clean and sustainable energy source for life support schemes and other set-up. Despite the use of green hydrogen in space investigation showing great potential it is essential to address the encounters associated with setup, i.e., storage, and transport of hydrogen in the unique environment of space. As space agencies and private corporations continue to update and collaborate, green hydrogen has the capacity to become an essential part of sustainable and cost-effective space examination endeavors.

8.5 ELECTRICITY GENERATION

The electricity production market via green hydrogen is a developing and innovative sector. This market has the capacity to change the energy scenery and provides numerous benefits for clean and reliable power generation. de Oliveira et al. [12] estimated the cost of electricity generated by green energy hydrogen. The authors assessed the electricity cost and its impact on the Iberian electricity market. The most significant features of the green energy hydrogen market in electricity generation are grid integration, distributed generation, hydrogen fuel cells, grid decarbonization, combined heat and power (CHP), off-grid and remote power, microgrids, energy storage, transitioning fossil fuel plants, environmental benefits, economic growth, technological policy and regulatory support, advancements, and global teamwork.

Green hydrogen can be utilized in hydrogen fuel cells to produce electricity. Fuel cells are highly effective and generate electricity with zero emissions, making them a smart option for a clean power cohort. Green hydrogen can be utilized to equate the grid by storing excess renewable energy. Hydrogen fuel cells can be utilized to generate electricity, confirming a stable and reliable power supply, especially during high electricity demand periods or when renewable energy sources are not available. This pays to efforts to decrease carbon releases in the energy sector. Green hydrogen technology permits supply generation, where electricity can be manufactured and stored locally, dipping transmission and distribution losses and improving grid resilience. Green hydrogen can offer off-grid and remote areas with a dependable source of electricity, including rural communities, islands, and remote industrial processes.

Hydrogen fuel cells can be utilized for joint heat and power uses, concurrently, producing electricity and useful heat. CHP systems are highly effective and can be utilized in commercial, residential, and industrial situations. Green hydrogen can be integrated into microgrid systems, which are small, restricted energy grids that can function individually or in combination with the central grid. This improves energy flexibility and security. Hydrogen can be utilized as a means of storing extra renewable energy, especially when the supply of wind or solar power surpasses the immediate mandate. The stored hydrogen can later be utilized to create electricity during high-demand periods. Green hydrogen can be utilized to retrofit current fossil fuel power plants, permitting them to decrease their carbon releases by substituting or co-firing with hydrogen. The application of green hydrogen in electricity cohort aids in reducing air pollution, combating climate change, and increasing air quality.

The progress of the green hydrogen electricity cohort market can motivate economic development, investment in research, development, and job creation. Ongoing research and development efforts are concentrated on improving the efficacy and inexpensiveness of hydrogen fuel cell technology for the electricity cohort. Governments are applying incentives and policies to encourage green hydrogen use in the electricity cohort, such as carbon pricing, and renewable energy standards. The electricity cohort market often includes international associations, sharing knowledge and resources to advance green hydrogen technology. The growing electricity manufacture market by green hydrogen can play an essential role in the change-over to cleaner, more sustainable, and more dependable power systems. However, problems like cost decrease, set-up development, and grid amalgamation need to be addressed for the full application of its potential.

8.6 DEVELOPMENT OF SOCIETY

The green hydrogen holds a good capacity for the development of society in numerous ways. Green hydrogen can donate to a more environmentally responsible, sustainable, and economically feasible future. Green hydrogen is predicted to play an important role in the development of society, with the capacity to reshape the world energy map and produce a USD 1.4 trillion market by 2050. The most significant aspects of the green energy hydrogen market in society development are dipping greenhouse gas emissions, energy security and individuality, economic development, clean transportation, industry decarbonization, job creation, grid decarbonization, energy access, technological advancements, infrastructure development, environmental benefits, rural development, public health policy and regulations, quality of life, and international collaboration [13].

The utilization of green hydrogen assists in reducing carbon emissions in numerous sectors, including transportation, industry, and electricity generation. This contributes to mitigating climate change and making a healthier environment for civilization. Green hydrogen can be manufactured locally, dipping a nation's reliance on energy imports. This improves energy security, stabilizes energy prices, and decreases vulnerability to geopolitical clashes over fossil fuel assets. Green hydrogen can be utilized in FCEVs, providing zero-emission transport options that improve air quality, decrease noise pollution, and support a transition to sustainable mobility. Green hydrogen can help reduce the carbon footprint of energy-intensive industries like chemicals, steel manufacturing, and petroleum refining. This is important for sustainable industrial growth. Green hydrogen can be utilized to store excess renewable energy and balance grid mandate. This assists the changeover to a cleaner and more reliable energy grid, helping society as a whole.

Green hydrogen's clean manufacture and use contribute to better air quality, minimum water usage, and a low ecological footprint compared to classical energy sources. Green hydrogen will have a positive influence on public health and the overall well-being of society just by decreasing air pollution and mitigating climate change. Governments play a vital role in fostering the green hydrogen industry by applying regulations and policies that support clean hydrogen manufacture, research and development, and its utilization in numerous sectors. The expansion of a green hydrogen infrastructure, including manufacturing facilities, transport networks, and refueling or recharging stations, builds investment chances and enables societal access to clean energy. Green hydrogen projects like solar installations and wind farms often deliver economic occasions for rural communities through lease payments to landowners and local tax income. Green hydrogen can spread access to dependable energy for remote and off-grid societies, refining quality of life, supporting education, and attractive healthcare services. The evolution of the green hydrogen industry inspires economic development and job creation, including positions related to production, construction, research and development, and operations.

The research and development in green hydrogen technologies drive invention and can lead to improvements in various areas, including energy storage, energy efficiency, and grid integration. The world nature of energy markets, generally, requires the international association to advance the green hydrogen industry. The collaboration among companies, countries, and research institutions can quicken the acceptance of this sustainable energy source. Briefly, the green hydrogen market has the possibility to be a key driver of sustainable growth, dipping greenhouse gas releases, generating jobs, augmenting energy security, and improving the quality of life for societies and individuals. Its acceptance is not only an ecological imperative but also a strategic and social one, as it provides a wide range of benefits that support the development of society.

8.7 CLIMATE ISSUES

The hydrogen market plays a noteworthy role in solving climate problems, mainly by providing a versatile and clean energy carrier that can be utilized in numerous sectors to decrease greenhouse gas releases and fight climate change [14–17]. The most significant features of the green energy hydrogen market in the climate problems are decarbonizing energy creation, carbon-neutral transport, decarbonizing heavy industry, clean heat and power cohort, energy store, climate flexibility, resource for carbon capture and store, rural and remote access, air quality enhancement, climate policy support, technological developments, international cooperation, carbon pricing, and climate mitigation and adaptation.

Specifically, green hydrogen is a carbon-free energy carrier that can substitute fossil fuels in many applications, considerably reducing CO_2 releases in the energy sector. Hydrogen fuel cells can power vehicles including cars, trucks, buses, and trains. This provides a clean and zero-emission substitute to internal combustion engines, dropping releases in the transport sector, which is a significant contributor to climate change. Hydrogen can be utilized as a reducing agent in industries like cement, steel, and chemicals, where conventional processes emit substantial carbon. The changeover to hydrogen can help decrease the carbon footprint of these energy-intensive segments. Hydrogen can be used in turbines or applied in fuel cells to create electricity and heat with zero greenhouse gas releases. This technology can aid in replacing coal and natural gas power plants, dipping releases from the electricity generation segment. Hydrogen can serve as an energy storage medium, taking excess energy from renewable sources when it is plentiful and liberating it when desired. This improves grid dependability and enables the incorporation of intermittent renewables like solar and wind. Hydrogen technology can improve the resilience of critical structures and remote societies by providing backup power during dangerous weather conditions and emergencies. Hydrogen manufacture and supply can extend clean energy access to remote and off-grid areas, enlightening life quality and fostering economic development. Hydrogen used in transport and industry results in zero releases of harmful contaminants, leading to improved air quality in urban regions [18].

The expansion of hydrogen technology and set-up can lead to inventions and advancements that advance energy efficacy and decrease the cost of hydrogen manufacture and use. Hydrogen supports climate policies and emissions decrease targets set by governments and international agreements, contributing to global efforts to fight climate change. Many nations are investing in hydrogen research, development, and international collaboration to improve the technology and meet global climate objectives. Hydrogen can benefit from carbon pricing mechanisms, such as cap-and-trade systems and carbon taxes, which incentivize the use of low-carbon and carbon-free energy sources. Hydrogen technology can support both climate justification efforts by dipping emissions and climate acceptance through improved energy resilience and dependability. The hydrogen market is a versatile tool in the contest against climate alteration, offering solutions that span numerous sectors and applications. However, to maximize its climate benefits, problems such as infrastructure development, cost reduction, and policy support must be addressed. As the hydrogen market grows and evolves, it has the future to make a noteworthy contribution to global efforts to decrease greenhouse gas releases and limit the influences of climate change.

8.8   IMPACT ON THE SOCIETY

Green hydrogen energy has a noteworthy impact on society in various ways. The most important features of green energy hydrogen on society are greenhouse gas emissions reduction, energy storage, decarbonization of key sectors, grid balancing, economic growth and job making, energy security, divergence of energy sources, technological developments, access to remote areas, decrease of water use, resilience and disaster retrieval, agriculture and food production, and international association. Green hydrogen is manufactured using renewable energy sources, like solar or wind power, that generate no direct greenhouse gas emissions during its manufacture. It can substitute fossil fuels in numerous sectors, dropping carbon emissions and mitigating climate alteration. Green hydrogen can be utilized in segments that are challenging to electrify directly, such as long-haul transportation, heavy industry, and aviation. It permits the decarbonization of these segments, dipping their environmental influence. Green hydrogen can serve as a means of storing excess renewable energy. It can be utilized to equilibrate the grid, confirming a steady and dependable energy supply, principally in areas with high shares of recurrent renewable energy sources. The growth and scaling of green hydrogen technologies can lead to job creation and economic progress. This includes industrial production, research and development, and the construction of production facilities.

Green hydrogen can augment energy security by reducing the necessity of fossil fuels. It provides a locally producible and sustainable energy source, which can assist in stabilizing energy markets. Green hydrogen expands the energy mix, dropping vulnerability to supply disturbances and price variations in the fossil fuel market. The growth of green hydrogen technologies drives revolution and can lead to progress in areas like renewable energy integration, electrolysis efficiency, and energy storage solutions. Green hydrogen can deliver access to clean energy for remote or off-grid communities, contributing to better quality of life, education, and healthcare. By reducing air pollution and reducing carbon emissions, green hydrogen can have a positive influence on public health, lowering the frequency of respiratory diseases and improving overall well-being. Green hydrogen manufacturing methods frequently use less water compared to some traditional power generation technologies, contributing to water preservation. The distributed green hydrogen manufacturing facilities and microgrids can augment community resilience during power outages and natural tragedies by providing backup power and helping disaster recovery efforts. Green hydrogen can be utilized in agriculture applications such as fertilizer production, which can help advance crop yields and contribute to food security.

The development of the green hydrogen transport sector can lead to job creation, economic growth, and investment in research and development of fuel cell technology.

Green hydrogen is fostering international association as nations work together on research, development, and placement to advance the technology and meet global sustainability objectives. Briefly, green hydrogen energy has the capability to change the energy landscape and play a significant role in the changeover to a more sustainable and low-carbon future. Its widespread acceptance can have far-reaching social benefits by dropping emissions, improving energy security, and driving economic growth and invention. By developing green energy hydrogen our society will be green without any pollution, environmental side effects, and serious diseases. However, it is essential to address challenges like cost decrease and infrastructure growth to fully unlock its potential.

REFERENCES

1. https://www.marketsandmarkets.com/Market-Reports/hydrogen-energy-storage-market-107179995.html
2. S. Atilhan, S. Park, M.M. El-Halwagi, M. Atilhan, M. Moore, R.B Nielsen, Green hydrogen as an alternative fuel for the shipping industry, *Curr. Opin. Chem. Eng.*, 31, 100668 (2021).
3. https://www.marketsandmarkets.com/Market-Reports/green-hydrogen-market-92444177.html
4. S. Cerniauskas, T. Grube, A. Praktiknjo, D. Stolten, M. Robinius, Future hydrogen markets for transportation and industry: The impact of CO_2 Taxes, *Energies*, 12, 4707 (2019).
5. A.M. Oliveira, R.R. Beswick, Y. Yan, A green hydrogen economy for a renewable energy society, *Curr. Opin. Chem. Eng.*, 33, 100701 (2021).
6. Y. Zheng, F. He, X. Shen, X. Jiang, Energy control strategy of fuel cell hybrid electric vehicle based on working conditions identification by least square support vector machine, *Energies*, 13, 426 (2020).
7. K. Song, Y. Wang, C. An, H. Xu, Y. Ding, Design and validation of energy management strategy for extended-range fuel cell electric vehicle using bond graph method, *Energies*, 14, 380 (2021).
8. A. Ajanovic, R. Haas, Prospects and impediments for hydrogen and fuel cell vehicles in the transport sector, *Int. J. Hydrogen Energy*, 46, 10049–10058 (2021).
9. The Future of Hydrogen: Seizing today's opportunities, Report prepared by the IEA for the G20, Japan, IEA Publications (2019).
10. O. Fakhreddine, Y. Gharbia, J.F. Derakhshandeh, A.M. Amer, Challenges and solutions of hydrogen fuel cells in transportation systems: A review and prospects, *World Electr. Veh. J.*, 14, 156 (2023).
11. A. Odenweller, F. Ueckerdt, G.F. Nemet, M. Jensterle, G. Luderer, Probabilistic feasibility space of scaling up green hydrogen supply, *Nat. Energy*, 7, 854–865 (2002).
12. A.R. de Oliveira, J.V. Collado, J.T. Saraiva, S. Doménech, F.A. Campos, Electricity cost of green hydrogen generation in the Iberian electricity market, *2021 IEEE Madrid PowerTech*, 28 June 2021–02 July 2021, Madrid, Spain (2022).
13. M.S. Ziegler, J.M. Mueller, G.D. Pereira, J. Song, M. Ferrara, Y.-M Chiang, J.E. Trancik, Storage requirements and costs of shaping renewable energy toward grid decarbonization, *Joule*, 3, 2134–2153 (2019).
14. S. Dunn, Hydrogen futures: Toward a sustainable energy system, *Int. J. Hydrogen Energy*, 27, 235–264 (2002).

15. P. Berrill, A. Arvesen, Y. Scholz, H.C. Gils, E. Hertwich, Environmental impacts of high penetration renewable energy scenarios for Europe, *Environ. Res. Lett.*, 11, 014012 (2016).

16. S. van Renssen, The hydrogen solution? *Nat. Clim. Change*, 10, 799–801 (2020).

17. A.I. Osman, N. Mehta, A.M. Elgarahy, M. Hefny, A. Al-Hinai, A.H. Al-Muhtaseb, D.W. Rooney, Hydrogen production, storage, utilisation and environmental impacts: A review, *Environ. Chem. Lett.*, 20, 153–188 (2022).

18. I.B. Ocko, S.P. Hamburg, Climate consequences of hydrogen emissions Atmos, *Chem. Phys.*, 22, 9349–9368 (2022).

9 Future challenges

9.1 INTRODUCTION

The advances in hydrogen manufacture, storage, and transportation technologies are required to make the process more effective and cost-effective. Creating a market for hydrogen-based products and services is essential for the success of hydrogen as an energy carrier. This includes developing applications like fuel cells for transportation, industrial processes, and power generation. Governments and regulatory bodies need to provide incentives and establish regulations that support hydrogen production, storage, distribution, and utilization, keeping the environmental safety concerns.

In spite of a great prospect for speeding up the energy transition, still green hydrogen recognizes economic, technical, and social challenges. There are many challenges and, in this chapter, these challenges are categorized as challenges in large-scale production, storage, transportation, and applications including the cost [1]. These problems should be overcome before green hydrogen can be manufactured and used. With regard to the technical issues, an important amount of energy is lost when green hydrogen is electrolyzed, liquefied, or changed to other carriers, distributed, and utilized in fuel cells. The most significant challenges are discussed below.

9.2 CHALLENGES IN LARGE-SCALE PRODUCTION

The large-scale manufacture of hydrogen energy faces numerous noteworthy issues. The most significant challenges are cost, poor infrastructure, low advanced technology, safety measures, environmental concerns, leakage, etc. Large-scale hydrogen manufacture poses the serious risk of explosions, fires, and detonations that can be even more disastrous than a detonation. This risk can be alleviated by understanding the hydrogen's flammability and considering it when planning hydrogen facilities. The experts should be registered to engineer amenities that avoid making high amounts of hydrogen in the atmosphere. These should be based on but also surpass the safety standards of natural gas amenities. Other emergency strategies, like well-defined spurt ways and breathing units, would also help alleviate the danger. Another noteworthy challenge is the extraordinary price of green hydrogen manufacture at a large scale. The existing manufacturing methods like electrolysis and steam methane reforming (SMR) are expensive and need noteworthy advancement in infrastructure. The manufacturing of low-carbon hydrogen by using either natural gas or renewables will need government action applying taxes or other mechanisms to fix the cost of carbon high adequate to warrant the cost-effectiveness with the mandatory, fossil-concentrated hydrogen and possibly supporting its manufacture to close the cost gap between hydrogen and other low-carbon fuels. Existing methods for hydrogen manufacture may not scale professionally to meet the need for

DOI: 10.1201/9781003432364-11

a hydrogen-based economy. Developing and optimizing large-scale manufacturing methods are essential to warrant a stable supply. Hydrogen can be manufactured using numerous feedstocks, including water, natural gas, and biomass. The accessibility and sustainability of these feedstocks require to be deliberated to ensure a long-term, environmentally friendly hydrogen distribution. Water and sunlight are free available materials for hydrogen production. The more advanced methods including photocatalysts and energy radiations should be used and applied to achieve large-scale hydrogen production with minimum cost [2].

Currently, most of the hydrogen is produced by SMR, which uses natural gas as a feedstock. This process produces greenhouse gases and is energy exhaustive. Developing low-energy input methods like electrolysis using renewable energy sources is crucial. While hydrogen is often seen as a clean energy carrier, the environmental influence of hydrogen manufacturing methods should be considered. For example, fossil fuels in hydrogen manufacture lead to carbon releases. Green hydrogen produced by renewable energy is more ecologically friendly but may face issues related to land use and resource obtainability for renewables [3]. Hydrogen is a leak-prone gas with an effective warming effect that is extensively overlooked. When hydrogen leaks into the atmosphere before it is used, it reacts with other chemicals to generate warming effects. The research has observed that, on time scales of a decade or two, hydrogen's heating power is much greater than before. This shows a contest for the industries. Because hydrogen molecules are tiny, they are prone to leakage. Therefore, minimizing leaks must be a primacy for every hydrogen project. Good regular inspections, engineering, and eliminating venting are critical.

The development of techniques and infrastructure in the hydrogen sector is at a slower step, hence impeding the widespread adoption of hydrogen. While the final cost for the customers depends severely on the efficacy and organization of manufacture as well as refueling set-up. The hydrogen supply is also attempted by national and local governments via a suitable arrangement and management. Briefly, large-scale manufacture of hydrogen energy has the talent to aid the world in moving to its net zero emissions goal. However, to meet this latent, hydrogen services and industries utilizing hydrogen and its derivatives should carefully evaluate the risks and fix the easing strategies to ensure safety standards. Furthermore, price, infrastructure, technology, and leakage are some of the issues that must be tackled to make hydrogen energy a feasible alternative to conventional fossil fuels.

9.3 CHALLENGES IN STORAGE

Hydrogen energy storage recognizes numerous challenges that need to be tackled to make it feasible and effective for storage at large scale [4,5]. Some of the key challenges are leakage, low energy density, and permeation, storage technologies, safety concerns, energy efficiency, stability, purity, cost, cycle life, infrastructure and compatibility, material availability, regulatory and safety standards, and environmental influence. Hydrogen can be stored in the form of a gas or liquid. The storage of hydrogen as a gas naturally needs high-pressure tanks (350–700 bar [5,000–10,000 psi] tank pressure). The storing of hydrogen as a liquid requires cryogenic temperatures because the boiling point of hydrogen at 1 atmosphere pressure is $-252.8°C$.

This makes hydrogen storage as a liquid more complex and costly. Hydrogen has a fairly low energy density by volume compared to classical fossil fuels. This means that a noteworthy volume of storage is needed to store the same quantity of energy, making it less practical for some applications. Hydrogen molecules are small and can permeate via many materials which can lead to leakage and loss of hydrogen stored. Developing materials with high hydrogen holding capacity is a challenge. Hydrogen is highly flammable, and safety concerns are of high importance. Preventing leaks, managing pressure, and ensuring the safety of storage systems are critical issues [6].

Several storage techniques are being explored, including compressed gas, liquid hydrogen, metal hydrides, and solid-state storage. Each of these techniques has its own challenges. The most significant issues are the need for specialized infrastructure and the development of effective, cost-effective materials. The storing process and retrieving hydrogen can result in energy losses, especially at high-pressure gas storage. Improving the storage energy efficacy of systems is a noteworthy challenge. For many storage techniques, the number of charging and discharging series a system can tolerate before degrading is a concern. Extending the storage system cycle life is important for long-term economic feasibility. Hydrogen storage systems are required to be scalable to fix the variable demands of dissimilar applications, from small-scale residential storing to large industrial and grid-scale stores.

The development of compatible hydrogen storage infrastructure with existing energy systems and transportation systems is a noteworthy challenge. This includes retrofitting standing facilities and evolving new ones. The stored hydrogen purity is significant for some applications, like fuel cells due to the chances of performance degradation. Maintaining high hydrogen purity during storage is a challenge. Some hydrogen storage techniques depend on specific materials, like critical metals or advanced composites. The availability of these materials in large quantities is a concern. The industry standards and regulations for hydrogen store systems for safety and reliability are essential. Industrialized and hydrogen storage materials disposal are challenging to handle from an environmental point of view. Therefore, the reduction of these influences and confirming the sustainability of store solutions is a challenge. The cost of hydrogen storage systems may be high, especially for advanced techniques like metal hydrides or solid-state storage. The reduction of the cost to make hydrogen storage economically feasible is important. Addressing these challenges will be important for the effective placement of hydrogen energy store systems that have the potential to play a precarious role in harmonizing renewable energy sources and permitting a clean and reliable energy distribution. Researchers, engineers, and policymakers are discovering advanced storage methods that have the capabilities for higher energy density and are cost-effective. Addressing the challenges associated with hydrogen storage is essential to realizing the full potential of hydrogen as a clean and sustainable energy carrier.

9.4 CHALLENGES IN TRANSPORTATION

The transportation of hydrogen energy presents numerous noteworthy challenges owing to the unique properties of hydrogen gas [7,8]. Some of the key challenges are leak and safety, infrastructure, energy intensity, materials compatibility, compression and liquefaction, range and distance, scaling infrastructure, cost, regulatory and safety standards, environmental influence, transportation modes, hydrogen purity, etc.

Hydrogen is highly flammable and has a low ignition energy. Ensuring the safe transport of hydrogen is a primary need. Any leaks during transport can result in accidental hazards. Ensuring the safety during transportation methods is critical. Developing and maintaining the essential infrastructure for transporting hydrogen are noteworthy challenges. These include tankers, pipelines, and storage facilities, all of which must be built to handle hydrogen's unique properties. Hydrogen can cause materials such as metals to become embrittled over time, possibly leading to leaks or structural failures. Ensuring that materials used in transportation systems are companionable with hydrogen is a concern. Hydrogen has a comparatively low energy density by volume compared to classical fuels, which means that a large volume of hydrogen must be transported to get the same energy content. This can result in energy losses during transportation [9].

Typically, hydrogen is transported as a compressed gas or liquid. However, liquefaction is a challenge and needs exhaustive energy. Hydrogen transport may be limited by the distance between production and consumption sites, as well as the choice of hydrogen fuel cell vehicles. The transportation distances need to be manageable to make hydrogen a practical option. The price of transporting hydrogen can be comparatively high, especially when considering the need for specialized infrastructure, safety measures, and energy losses during compression or liquefaction. Reducing transportation costs is important for the economic feasibility of hydrogen. Scaling up the infrastructure for distribution becomes a challenge as the use of hydrogen for transportation grows. Building and maintaining a network of refueling stations and transportation passages is an expensive and complex procedure. The safety and regulatory standards are essential for the transportation and storage of hydrogen. However, the development and harmonization of these standards are challenging. The transportation of hydrogen, especially when it involves compression or liquefaction, can have environmental effects. The reduction in the carbon footprint of the transportation process is also a concern [10,11].

The selection of the transport mode is very important by considering the economy and feasibility of manufacture and supply of the hydrogen. The most common modes of hydrogen distribution are pipelines, trucks, trains, and ships. These are affected by various factors such as distance, volume, and safety. These have their own types of challenges. Also, the purity of hydrogen is very challenging to maintain throughout its supply [12]. These challenges are crucial for the successful mixing of hydrogen into the transportation sector. Hydrogen transportation has to play a crucial role in decarbonizing various industries, but overcoming these difficulties will require partnership among government, stakeholders, industry, and research institutions.

9.5   CHALLENGES IN APPLICATIONS

The final destination of green energy hydrogen is its utility and applications. It will not be accepted by the public until all the applications are perfect. Green energy hydrogen has a wide range of applications and, consequently, every application has challenges. In this way, this section describes the maximum challenges of green energy hydrogen. Therefore, the applications of hydrogen energy recognize numerous challenges, depending on the precise use case. Some challenges linked with diverse applications of hydrogen energy are cost, infrastructure, transportation, limited vehicle accessibility, competing techniques, competing technologies, efficacy, carbon emissions, safety and compatibility, supply reliability, heating and cooling, and energy storage. [13].

The cost of hydrogen refueling stations is quite high and needs substantial investment. The number of hydrogen fuel cell vehicles on the market is relatively small, and limiting consumer choice, which gives a moderate amount of business. Hydrogen recognizes competition from electric vehicles, which have seen good market acceptance, especially in passenger cars. The traditional hydrogen combustion in power generation is less capable compared to other energy sources. If hydrogen is manufactured using fossil fuels, carbon emissions can still be a problem, making it essential to focus on "green" hydrogen manufactured using renewable energy sources. The integration of hydrogen into the existing power grid can be exciting and may need noteworthy modifications. The industries should make sure that hydrogen can be safely combined into existing processes and equipment, which may not be planned for hydrogen use. The dependence on a reliable supply of hydrogen is essential for continuous industrial processes. The development and deployment of hydrogen distribution networks for residential and commercial heating and cooling are noteworthy challenges.

Hydrogen storage and retrieval systems can face energy losses, mainly during compression and liquefaction processes. It is a critical concern to ensure the safe storage and release of hydrogen, especially for residential and commercial applications. Fuel cells need high-purity hydrogen, which may be a challenge to maintain during storage and transport. It is well known that hydrogen is used in ammonia manufacture. Ammonia is being used in various industries and, consequently, the costly production of ammonia (depends on hydrogen) is a challenge. Also, hydrogen cost needs to be low as maximum as possible when hydrogen is used as fuel for heating buildings. Hydrogen is used as fuel in rockets and launching and handling hydrogen in space applications is a challenge from the cost point of view. Hydrogen production cost is acceptable by using alternative sources of energy in hydrogen production. Therefore, hydrogen production is inexpensive in countries that have high amounts of alternative energy sources. However, the transportation of hydrogen from one country to another is challenging from an infrastructure point of view. Also, the challenges of safety and purity exist when transportation is carried out for long distances. In view of these facts, the concerted effort from governments, industries, and research organizations to advance the development, infrastructure, and safety measures associated with hydrogen energy should be addressed. Additionally, the continued growth of "green" hydrogen manufacture, using renewable energy sources, is critical for mitigating environmental issues and increasing the sustainability of hydrogen applications.

9.6 SOLUTION OF THE CHALLENGES

Hydrogen manufacturing faces several challenges as discussed above. These issues need to be tackled to make a sustainable and feasible energy source. Hydrogen distribution and consumption rates should be carefully matched, and regional or even local hydrogen production can maximize the use of local resources and minimize supply costs. All the challenges should be resolved in such a way that the green energy hydrogen may be human-friendly and tolerable to the public.

The cost of green hydrogen of high purity manufacture is a significant challenge. The electrolysis and other clean production methods are still more costly than fossil fuel-based hydrogen. Electrolysis can cause energy losses during the process. Hydrogen manufacturing methods should be more efficient. Research and development should focus on improving the efficiency of hydrogen production technologies. Therefore, we need to explore new methods to produce high-quality green hydrogen at low prices. Direct photo water splitting is the only choice but it needs more advanced research. Cost-effective and safe storage and transportation solutions are also needed after producing cost-effective hydrogen. These tasks may be achieved by advanced tanks and pipelines. But these tasks are challenging, especially the building of the essential infrastructure for hydrogen production, distribution, and utilization. These include creating a network of refueling stations for hydrogen fuel cell vehicles and pipelines for industrial use. Government and industry association is essential to develop this infrastructure [14].

Hydrogen production needs to be scaled up significantly to meet the growing demand for clean energy. The economic hydrogen manufacture at a large scale is necessary. Sometimes, electrolysis relies on critical materials like platinum for catalysts, which can have environmental and supply chain consequences. The safety of the hydrogen supply may be ensured by making high-quality containers and pipelines. Therefore, continuous research and development are required to improve existing hydrogen manufacturing methods and discover new, more effective, and sustainable methods. Innovations in catalysts, materials, and process optimization are urgently needed [15,16].

Governments must create supportive policies and regulations to inspire the acceptance of hydrogen as an energy carrier. The subsidies, incentives, and carbon pricing mechanisms should be stressed to encourage hydrogen manufacture and utilization. All these challenges may be addressed properly by the collaboration among governments, industries, and research institutions. It also necessitates a long-term commitment to sustainable and clean hydrogen production, as hydrogen has the potential to play a crucial role in decarbonizing various sectors of the economy, such as transportation and industry.

REFERENCES

1. F. Eljack, M.-K. Kazi, Prospects and challenges of green hydrogen economy via multi-sector global symbiosis in Qatar, *Front. Sustain.*, 1, 612762 (2021).
2. C.B. Agaton, K.I.T. Batac, E.M. Reyes Jr., Prospects and challenges for green hydrogen production and utilization in the Philippines, *Int. J. Hydrogen Energy*, 47, 17859–17870 (2022).
3. L.M. Eh Christina, A.N.T. Tiong, J. Kansedo, C.H. Lim, B.S. How, W.P.Q. Ng, Circular hydrogen economy and its challenges, *Chem. Eng. Trans.*, 94, 1273–1278 (2022).
4. N.N. Nguyen, Prospect and challenges of hydrate-based hydrogen storage in the low-carbon future, *Energy Fuels*, 37 (14), 9771–9789 (2023).

5. D. Mori, K. Hirose, Recent challenges of hydrogen storage technologies for fuel cell vehicles, *Int. J. Hydrogen Energy*, 34, 4569–4574 (2009).

6. U.S. Meda, N. Bhat, A. Pandey, K.N. Subramanya, M.A.L.A. Raj, Challenges associated with hydrogen storage systems due to the hydrogen embrittlement of high strength steels, *Int. J. Hydrogen Energy*, 48, 17894–17913 (2023).

7. R. Gerboni, Introduction to hydrogen transportation, In R.B. Gupta, A. Basile, T. Nejat Veziroğlu (Edts.), *Compendium of Hydrogen Energy, Volume 2, Hydrogen Storage, Distribution and Infrastructure*, Woodhead Publishing (2015).

8. M.S. Ali, M.S.H. Khan, R.A. Tuhin, M.A. Kabir, A.K. Azad, O. Farrok, Hydrogen Energy Conversion and Management, Hydrogen energy storage and transportation challenges: A review of recent advances, In M. Masud, K. Khan, A.K. Azad, A.M.T. Oo (Edts.), *Hydrogen Energy Conversion and Management*, 255–287, Elsevier (2023).

9. O. Fakhreddine, Y. Gharbia, J.F. Derakhshandeh, A.M. Amer, Challenges and solutions of hydrogen fuel cells in transportation systems: A Review and prospects, *World Electr. Veh. J.*, 14, 156 (2023).

10. A.M. Abdalla, S. Hossain, O.B. Nisfindy, A.T. Azad, M.M.K. Dawood, A. Azad, Hydrogen production, storage, transportation and key challenges with applications: A review, *Energy Convers. Manage.*, 165, 602–627 (2018).

11. N. Shakya, R. Shrestha, R. Saiju, B.S. Thapa, Hydrogen as a fuel for electrifying transportation sector in Nepal: Opportunities and challenges, *Conf. Ser.: Earth Environ. Sci.*, 1037, 012064 (2022).

12. S.A. Gorji, Challenges and opportunities in green hydrogen supply chain through metaheuristic optimization, *J. Comput. Design Eng.*, 10, 1143–1157 (2023).

13. L. Laffineur, R. Campe, P. Perreault, S.W. Verbruggen, S. Lenaerts, Challenges in the use of hydrogen for maritime applications Laurens Van Hoecke, *Energy Environ. Sci.*, 14, 815–843 (2021).

14. B. Zhang, S.-X. Zhang, R. Yao, Y.-H. Wu, J.-S. Qiu, Progress and prospects of hydrogen production: Opportunities and challenges, *J. Electro. Sci. Technol.*, 19, 100080 (2021).

15. A.I. Osman, N. Mehta, A.M. Elgarahy, M. Hefny, A. AlHinai, A.H. AlMuhtaseb, D.W. Rooney, Hydrogen production, storage, utilisation and environmental impacts: A review, *Environ. Chem. Lett.*, 20, 153–188 (2022).

16. E.B. Anderson, L.M. Moulthrop, K.E. Ayers, Hydrogen infrastructure challenges and solutions, *ECS Transact.*, 41, 75–83 (2012).

10 Future perspectives

10.1 INTRODUCTION

Hydrogen is emerging as a talented alternative to classical fossil fuels. This has the potential to be the green energy source of the future. The pressing requirement to eradicate carbon dioxide emissions, and harmful fossil fuels and a sustainable environment are driving the global green hydrogen boom. Green hydrogen technology is becoming more competitive in terms of cost, and the advanced development of the hydrogen ecosystem. This is growing in the areas on both the supply and demand sides, which is contributing to the growth of green hydrogen. Green hydrogen has the capability to meet around 12% of the world's energy mandate by 2050 to hit the Paris climate target. The costs of construction of a hydrogen infrastructure over the coming decades could run into the tens of trillions of dollars. Furthermore, the investment in hydrogen production, transport, and consumption will have to be undertaken instantaneously. The green hydrogen could become cost-competitive by 2030 as economies of scale drive down the cost of electrolyzers and the cost of wind and solar power continues to fall [1–3].

10.2 FUTURE OF HYDROGEN

The future of green hydrogen, often termed as the most sustainable form of hydrogen, holds significant potential. It is a central element of global efforts to transition to a low-carbon and environmentally responsible energy system. Some of the key factors for the future of green energy hydrogen are cost reduction, decarbonization, international trade, renewable energy integration, hydrogen infrastructure, transportation, industrial applications, energy storage, and technological advancements.

The green hydrogen is expected to be a main driver in the global effort to decarbonize energy systems. It can substitute fossil fuels in various applications, from transportation to industrial processes, considerably reducing carbon emissions. As the share of renewable energy sources like solar and wind continues to grow, the capability to produce green hydrogen will expand. One of the most noteworthy challenges for green hydrogen is cost affordability. The technical advancements, economies of scale, and falling renewable energy costs are predictable to reduce the cost. The investment in hydrogen infrastructure, including storage, pipelines, and transportation, will be important for the widespread distribution and use of green hydrogen. Some regions, mainly those with ample renewable energy resources, are positioning themselves as potential green hydrogen exporters, creating a global hydrogen market and reshaping the energy trade.

Green hydrogen is likely to have a role in decarbonizing numerous modes of transportation, including heavy-duty and long-range applications in the automotive, aviation, shipping, and rail sectors. Many industries like steel and chemical

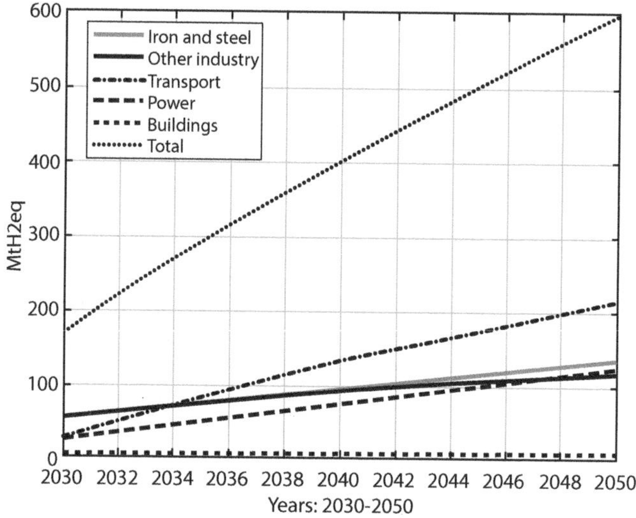

FIGURE 10.1 Future projection of green hydrogen demand in various sectors [reproduced with permission from ref. 3].

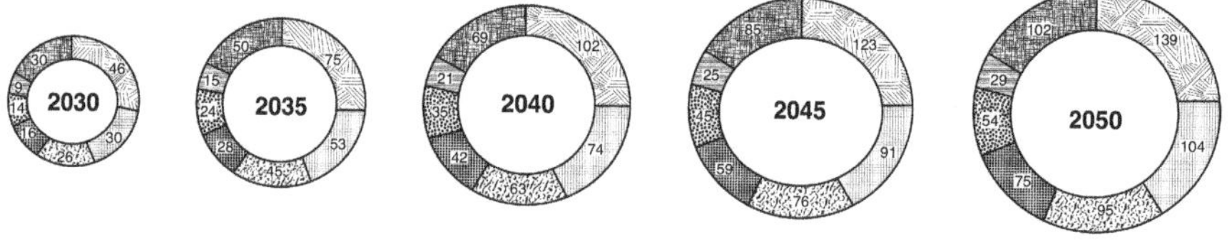

FIGURE 10.2 The regional demand for clean hydrogen and its derivatives (2,030–2,050 MtH$_2$eq) [reproduced with permission from ref. 4].

manufacturing are discovering the use of green hydrogen as a clean energy source for their methods. Green hydrogen can serve as a form of long-term energy storage, assisting to balance the intermittency of renewable energy sources and sustenance grid steadiness. The ongoing research and development in materials science, electrolysis technology, and storage systems will lead to developments in efficacy and cost-effectiveness.

Reducing the environmental influence of green hydrogen manufacture, mainly in terms of water use, land use, and emissions, is an attention for sustainable growth. Government policies and incentives, including tax credits, subsidies, and regulations, will play a noteworthy role in fostering the adoption of green hydrogen techniques and creating a favorable investment climate. The adoption of green hydrogen depends on the readiness of consumers and industries to embrace this technology. Raising awareness, ensuring acceptance, and stimulating market demand will be critical. The future of green hydrogen is tightly associated with the broader energy

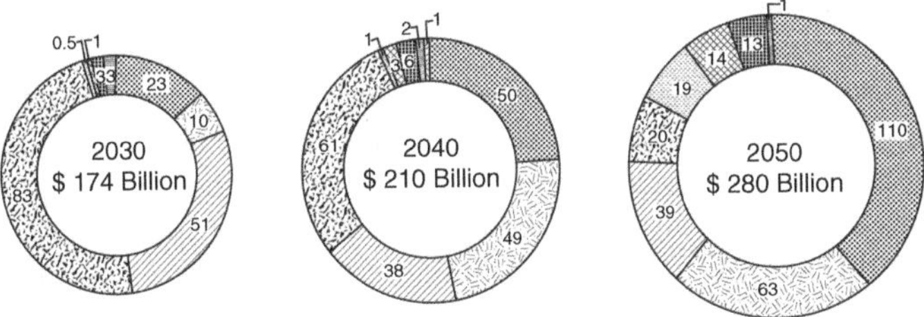

FIGURE 10.3   The annual export revenue (US$ billion) from 2030 to 2050 [reproduced with permission from ref. 4].

transition and the global promise to combat climate change. It has the potential to be a cornerstone of a cleaner, more sustainable, and resilient energy system. If the challenges of cost, infrastructure development, and policy support are addressed, green hydrogen could expressively contribute to a more sustainable and low-carbon energy future. The future demand for green hydrogen is shown in Figure 10.1 [3].

The regional future demand for green energy is shown in Figure 10.2. This figure clearly shows that the demand for green hydrogen energy is increasing continuously. The annual export revenue is also estimated to increase by 2050 (Figure 10.3) [4].

10.3   FUTURE OF WATER SPLITTING FOR HYDROGEN

The future of water splitting for hydrogen production holds substantial promise, especially as the world seeks to transition to cleaner and more sustainable energy sources. Water splitting can be a green and sustainable way to generate hydrogen, and advancements in technology and augmented renewable energy integration are key drivers for its future. Some of the key features of the future of water splitting for hydrogen production are cost reduction, renewable energy integration, electrolysis technologies, efficiency improvement, scaling up, hydrogen infrastructure, hydrogen storage, grid integration, environmental considerations, hydrogen applications, policy support, international collaboration, and market growth.

Water splitting using electricity from renewable sources, such as solar and wind, is expected to become progressively prominent. As the share of renewables in the energy mix rises, the capacity for green hydrogen production through water splitting will increase. The continued research and development will lead to developments in electrolysis technologies, making them more efficient and cost-effective. Both proton exchange membrane and alkaline electrolyzers are expected to see technological developments. Achieving cost competitiveness with other hydrogen production methods, such as natural gas reforming, is a crucial goal. Technological innovation and economies of scale will contribute to cost decrease. Enhancing the energy efficiency of water-splitting processes is a key area of research. More effective systems will decrease energy input requirements, making the process more economically and environmentally viable.

Many water-splitting projects are presently in the demonstration or pilot phase. The future will involve scaling up these projects to commercial and industrial levels.

The investment in infrastructure, such as storage facilities, pipelines, and distribution networks, is essential to enable the transportation and utilization of hydrogen produced through water splitting. The efficient and cost-effective hydrogen storage techniques, both for transportation and stationary applications, will be critical. Incorporating hydrogen manufacture through water splitting into the existing energy grid and establishing a hydrogen infrastructure that aligns with grid operations will require careful planning. The efforts to reduce the environmental impact of water splitting, especially in terms of water use and emissions, will be a focal point for sustainable growth. The growth of industries and applications utilizing hydrogen produced through water splitting, such as power generation, industrial processes, transportation, and heating, will be a significant part of the future.

Water splitting in the presence of sunlight with a suitable photocatalyst may be the most inexpensive technology in the future. The splitting of seawater could provide an infinite source of green hydrogen that could power heavy automobiles and decarbonize industries like steel and cement. The photo-electrochemical (PEC) water splitting is a gifted solar-to-hydrogen pathway, contributing to the potential for high conversion efficacy at low operating temperatures utilizing cost-effective materials. PEC water splitting utilizes semiconductor materials to change solar energy directly to chemical energy in the form of hydrogen. The recent advances in effective and scalable solar hydrogen manufacture through water splitting have been made, and the challenges facing the industrialization of hydrogen manufacture from solar water splitting are being examined [5–9]. The future development of smart photocatalysts will be a boon for water splitting [10]. The most important photo-water splitting methods are photocatalytic, PEC, and photovoltaic-electrochemical (PV-EC). The different ways of photo-water splitting are shown in Figure 10.4.

The governments of different countries are supposed to play a critical role in supporting water-splitting methods via tax credits, subsidies, and regulatory measures. International collaboration on research, development, and deployment of water-splitting techniques will help harness the global potential for hydrogen manufacture. The building market for hydrogen and hydrogen-based products, both domestically and through international trades, is supposed to be a key driver of future technology. The future of water splitting for hydrogen manufacture is closely related to the broader energy transition and the world's promise to mitigate climate change. With growing advancements in technology, cost reductions, and increased policy support, water splitting has the potential to play an important role in a more sustainable and low-carbon energy future.

10.4 FUTURE GREEN WORLD

Green hydrogen energy has the talent to play an important role in the future of the world's energy supply. Green hydrogen is a talented substitute to traditional fossil fuels, and it has the prospective to be a key component of the green energy future. The green hydrogen energy has the talent to shape the world in a more sustainable and environmentally responsible way. Green hydrogen could become cost-competitive by 2030 as economies of scale drive down the price of electrolyzers and the cost of solar and wind power endures to fall. The trend of replacement of natural gas by

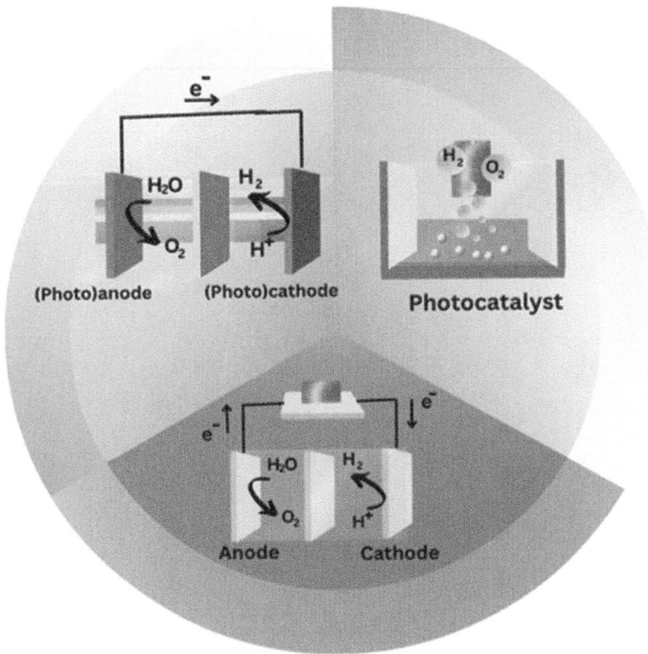

FIGURE 10.4 Different methods of photo-water splitting.

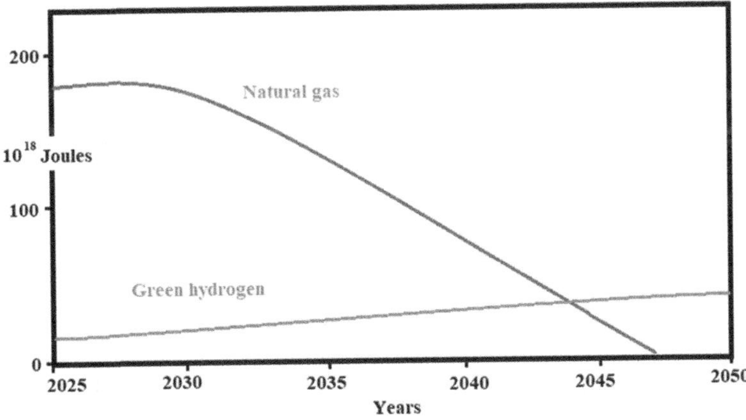

FIGURE 10.5 A trend of replacement of natural gas by hydrogen.

hydrogen is shown in Figure 10.5. This figure dictates that the natural gas supply will decrease rapidly till 2045 while the total hydrogen demand will increase.

Replacing fossil fuels with green hydrogen will radically decrease emissions from transportation and industries. The green hydrogen could be the fuel of the future. It has the talent to become a significant driver of a worldwide green energy change.

FIGURE 10.6 Future speculation of the world due to green energy hydrogen.

Briefly, green hydrogen energy has the talent to shape the world in a more sustainable and environmentally probable way. The time is right to tap into hydrogen's talent to play a key role in a secure, clean, and affordable energy future. The pressing need to remove harmful fossil fuels and carbon dioxide emissions for a more sustainable, environmentally responsible future is driving the global green hydrogen boom. The green hydrogen projects, while hopeful, still grapple with technological and geopolitical uncertainties, as well as financial constraints and environmental concerns. Finally, we can assume that the world will be heaven with no pollution, good health, and plenty of jobs. Everyone will enjoy and a speculation of the world is shown in Figure 10.6.

REFERENCES

1. A. Pareek, R. Dom, J. Gupta, J. Chandran, V. Adepu, P.H. Borse, Insights into renewable hydrogen energy: Recent advances and prospects, *Mater. Sci. Energy Technol.*, 3, 319–327 (2020).
2. R.V. Vardhan, R. Mahalakshmi, R. Anand, A. Mohanty, A review on green hydrogen: Future of green hydrogen in India, In *2022 6th International Conference on Devices, Circuits and Systems (ICDCS)*, 21–22 April 2022, Coimbatore, India (2022).
3. I. Marouani, T. Guesmi, B.M. Alshammari, K. Alqunun, A. Alzamil, M. Alturki, H.H. Abdallah, Integration of renewable-energy-based green hydrogen into the energy future, *Processes*, 11, 2685 (2023).
4. B. Lorentz, J. Trüby, F.C. Matthes, P. Philip, Green hydrogen: Energizing the path to net zero, Deloitte's 2023 global green hydrogen outlook (2023); https://www.deloitte.com/global/en/issues/climate/green-hydrogen.html
5. R.M.N. Yerga, M.C.A. Galván, F. del Valle, J.A.V. de la Mano, J.L.G. Fierro, Water splitting on semiconductor catalysts under visible-light irradiation, *ChemSusChem*, 2, 471–485 (2009).

6. C.-H. Liao, C.-W. Huang, J.C.S. Wu, Hydrogen production from semiconductor-based photocatalysis via water splitting, *Catalysts*, 2, 490–516 (2012).
7. H. Song, S. Luo, H. Huang, B. Deng, J. Ye, Solar-driven hydrogen production: Recent advances, challenges, and future perspectives, *ACS Energy Lett.*, 7, 1043–1065 (2022).
8. P. Hota, A. Das, D.K. Maiti, A short review on generation of green fuel hydrogen through water splitting, *Int. J. Hydrogen Energy*, 48, 523–541 (2023).
9. Y. Zheng, M. Ma, H. Shao, Recent advances in efficient and scalable solar hydrogen production through water splitting, *Carbon Neutrality*, 2, 23 (2023).
10. S. Wang, A. Lu, C.J. Zhong, Hydrogen production from water electrolysis: Role of catalysts, *Nano Converg.*, 8, 4 (2021).

Index